AF522429

Information Technology and E-Governance

Information Technology and E-Governance

Jaskiran Kaur

Information Technology and E-Governance

ISBN 978-93-5111-505-2

Published in 2015 in India by

RANDOM PUBLICATIONS

4376-A/4B, Gali Murari Lal, Ansari Road

New Delhi-110 002

Phone : +9111-43580356, 011-23289044, 011-43142548

e-mail: sales@randompublications.com,

info@randompublications.com, randomexports@gmail.com

Type Setting by : Friends Media, Delhi-110089

Printed at : **Thomson Press (India) Ltd.**

Preface

E-Governance is the use of a range of modern Information and Communication Technologies such as internet, local area networks, mobiles etc. by Government to improve the effectiveness, efficiency, service delivery and to promote democracy. E-Government can transform citizen service, provide access to information to empower citizens, enable their participation in government and enhance citizen economic and social opportunities, so that they can make better lives, for themselves and for the next generation.

Countries around the world have realized the importance of electronic government or e-Government and the impact technology has in transforming the lives of their citizens through the efficient and judicious use of information technology in the Government sector.

E-Governance is the future, many countries are looking forward to for a corruption free government. The essence of E-governance is to reach the beneficiary and ensure that the services intended to reach the desired individual has been met with. There should be an auto-response system to support the essence of E-governance, whereby the Government realizes the efficacy of its governance. E-governance is by the governed, for the governed and of the governed.

The present book is a comprehensive, well-balanced and up-to-date resource on information technology and e-governance. It provides policy information on e-democracy, access issues, privacy, security, regulatory, enforcement and taxation issues, as well as management information on business plans, public-private partnerships, strategic planning, project management, implementation factors, and evaluation. The book is intended to serve as a useful guide to IT consultants, government officials, students, educationalists, NGOs and public in general.

Author

Contents

1

Integrating Information Technology into Public Administration

Practitioners of public administration continue to debate the extent to which public institutions and organizations are likely to be transformed by the burgeoning adoption of information technology (IT). Scholars who emphasize the concept and impact of "informatization" are among those who believe that there will be a substantial, even a revolutionary, transformation. Interpreted broadly, this term refers to the process by which "public service organisations - and thus the administrative organisation of the state - are becoming strategically and centrally dependent upon the changing flows of informational resources which are made possible by powerful combinations of information and communication technologies."

Concept of Informatization

There has been little scholarly examination of the concept of "informatization." The concept is closely associated with academic scholarship in Western Europe and, most notably, with the Rotterdam/ Tilburg research group on informatization in public administration, which explains the process of informatization as:

- the introduction of information technology to shape or take care of the information retrieval process by means of automated information systems;

- the (re)arrangement of information flows and information relationships to facilitate the administrative or management information process;
- the adjustment or change of the organizational structure in which information technology is introduced;
- the development of information policies as a differentiated area of decision-making in the organization; and
- the introduction of specific expertise in the field of information technology through functionaries or consultants with specific tasks in this field.

Some academic scholars view informatization as an emerging new paradigm in public administration. John Taylor argues that this perspective enables us to understand better than we ever have the relationships between governments and citizens, political executives and legislators, and political leaders and public servants. He goes so far as to recommend a breaking away from mainstream writings on public administration, since these writings ignore or minimize the fact that information is "the centrepiece" of understanding public administration" and "from that understanding, we can research and understand public administration at a level that has previously been impossible to achieve."

A. Zuurmond and I.Th.M. Snellen go even further. They contend that "[facilitated by [information and communication technologies]-developments and attracted by iCT-opportunities, bureaucracies are being replaced by infocracies." Infocracy is offered as a new ideal-type that is fundamentally different from the traditional Weberian ideal-type bureaucratic structure. This new organizational structure is characterized by

- less hierarchy (because of the disappearance of the middle management layer);
- greater decentralization (because team members are empowered);
- less formalization (in that forms are shorter and approvals are fewer);
- a change in the level of specialization (in that the scope of public servants' tasks has broadened but the extent to which they are in charge of how their work is done has shrunk); and
- two forms of change in specialization (a new official - the IT expert - has been introduced, and there has been a de-skilling because fewer officials with high vocational training are needed).

While there is substantial agreement among academic authors on informatization that information is the lifeblood of public administration, there is disagreement about the importance of *information technology* in the informatization process. Christine Bellamy and John Taylor offer a comparatively restrained assessment of the likely impact of the use of ICTS. They assert that "information and its communication cannot... be corralled into a technologically defined pen, to be pigeonholed, studied and researched under the rubric of IT." They conclude also, from their application of institutional analysis to information-age governance, that "the institutions of governance will mould and fashion the *revolutionary potential* of ICTS into an *evolutionary reality.*"

Similarly, John Hudson's analysis of informatization in the light of some leading theoretical approaches in political science suggests that ITS will have a much more limited effect than that envisaged by widespread Utopian perspectives on the role of IT. His use of the policy networks, rational choice and new institutionalism approaches to examine informatization in three policy sectors of British government led him to conclude that

- change has been incremental in nature;
- there were variations among sectors in the extent of informatization; and
- these variations were significantly influenced by political considerations.

This chapter does not argue that informatization has created a new paradigm of public administration or a new ideal-type of public bureaucracy. It does argue, however, that the current and emerging impact of IT on the public sector has been underestimated, not only by many practitioners but by many public administration scholars as well. Most scholarly publications on public administration in Canada, including the most recent textbooks, have little or nothing to say about the expansion and implications of the use of IT. However, in *The New Public Organization,* Kenneth Kernaghan, Brian Marson and Sandford Borins emphasize the growing power and pervasiveness of IT in public organizations. They conceptualize IT as one of the primary means by which public organizations are moving between the bureaucratic and the post-bureaucratic models of public organization. They argue that IT "is right at the core of public organizations in Canada" and when IT "is introduced into the public sector, it can dramatically affect both internal operations and relations with the public." Academic scholars need to take careful account of the impact of IT, not only on the structures and management of government but also on theories of public organization (e.g., Weberian ideal-type bureaucracy, open systems organization theory).

The analysis in this chapter lends support to those who foresee increasing reliance on IT as bringing about more substantial change in public organizations than that envisaged by many theorists and practitioners of public administration. We argue that the use of IT in public organizations tends to lead to certain kinds of changes in these organizations and that the more extensive the usage the more likely these changes are to occur. However, it is acknowledged that the nature and extent of these changes can be greatly influenced not only by technological forces but by political and other forces as well. Kernaghan, Marson and Borins acknowledge that IT is only one of several major drivers of public-service reform (e.g., restructuring, partnerships). Moreover, the use of each driver has important political, organizational, managerial and value implications, and there are substantial impediments to the successful application of each driver, including IT.

Information Technology as a Resource

A focus on financial and human resources management continues to dominate general management concerns in the public sector. However, the relative importance of information management (IM) is increasing at an accelerated pace - and managing IT is an integral part of the broad task of IM in general. The elevated importance of IM - and of IT as well - is formally acknowledged in the federal government's 2003 *Policy on the Management of Government Information.* This policy describes information in the digital age as a valuable asset that the government must manage as a public trust. The policy also requires federal government institutions to "use electronic systems as *the preferred means* of creating, using and managing information." Information technology will be the central resource in meeting this requirement.

Information and information technology are both becoming increasingly important public-sector resources and they are tightly intertwined. They can be viewed as commodities that are bought and sold in the marketplace, but they can also be viewed, in the organizational context, as resources that are vital to the daily tasks of individuals within the organization and consequently to its overall ability to meet its objectives.

Information is a less tangible resource than most physical goods and consumer commodities and it means different things to different managers. Its many forms include business data, voice conversations, still images, motion pictures, and multimedia presentations. A processing manager might conceive of information in terms of data, a records manager in terms of records and reports, and a senior manager in terms of conversations and written documents. Information technology is a more tangible resource than information; it refers to any equipment or interconnected system of equipment that includes all forms of technology used to create, store, manipulate, manage, move, display, switch, interchange, transmit or receive information in its various forms.

Contemporary IT allows information to be exploited to a far greater degree than older technological and material-based alternatives. This fact underpins the shift from the industrial to the information economy. In the industrial economy, the methods of production are material-based and resource-intensive; they consume considerable amounts of time, labour, capital and physical materials. In the information economy, the methods of production are not centred on "hard" resources but rather on the availability of information. Production processes consume less time and energy and require a more highly skilled workforce.

Shoshana Zuboff provides an important insight in her distinction between IT and traditional forms of machine technology. She argues that the use of IT permits not only automation but also what she terms "informating." The same technology that automates operations also informates by "simultaneously generating information about the underlying productive and administrative processes through which an organization accomplishes its work." This information can be used for a variety of purposes, including empowering employees, enabling management to exercise greater control over employees, and improving service. Zuboff notes the potential for innovation arising from intended rather than unintended use of the informating power of IT. She notes also that managers have not recognized sufficiently the power of informating and that they could choose at the start of an automation process to exploit this power.

Public administration is an information-intensive activity. Indeed, information is an indispensable resource that is both the raw material for, and a product of, virtually all government activities. Such activities as making health or defence policy, collecting taxes, providing social security services and conducting medical research are grounded in information. Moreover, the policy-making, decision-making and communication processes that pervade government's service, regulatory and research functions rely heavily on information. The transfer of information that occurs in these processes entails information flows between and among all of the many actors in the political system.

These information flows involve the collection, production, dissemination and exchange of information - and IT is the resource that is enabling these flows to take place with increasing ease, rapidity and repercussions. Information technology is the driving force of the informatization process. The infrastructure provided by increasingly sophisticated and powerful IT is making information more usable and valuable. The value of the World Wide Web, for example, lies in its capacity to provide immediate and widespread access to information.

As a resource like IT becomes increasingly preferred and prominent in an organization - a development that has been formally mandated by the federal government's *Policy on the Management of Government Information* noted above - it influences virtually all aspects of its environment. The extent of this influence on the organization and on the organization's consequent transformation depends to a substantial degree on the nature of the resource itself. Information technology is a resource with certain inherent predispositions. When a public organization uses IT extensively, these predispositions tend to lead to structural and managerial change.

Information Technology Predispositions

A strong argument can be made for three major predispositions:

1. *The use of IT predisposes public organizations to acquire specialized skills and substantial capital investment.* All resources in their barest form require some cultivation to make them usable. However, in its requirements for cultivation, IT differs from natural resources like iron ore or lumber, which require substantial energy and physical labour to turn them into something useful. The development of IT, in public organizations as elsewhere, is expensive and requires highly specialized employees and considerable investment in technological infrastructure. Governments must have substantial human and financial resources to ensure the effective formulation and implementation of IT policy.

2. *The use of IT predisposes public organizations to improve efficiency by streamlining the management of information.* Before the introduction of IT, the management of information resources was tied to paper-based processes that were energy and labour intensive. The use of computing technology significantly increases productivity by greatly enhancing the efficiency of information processing. In the information-rich environment of the public sector, this can lead to substantial cost-savings resulting from the elimination of organizational redundancy, including jobs.
3. *The use of IT predisposes public organizations to share and disseminate information.* Modern communications capabilities allow information to be transferred at close to the speed of light to countless individuals scattered around the globe. In sharp contrast, such material goods as books or pamphlets are more difficult to transport and cannot be disseminated as readily as a piece of information. The availability of personal desktops and other ITS has made communication between organizational members simple and effective, as measured by cost and energy expended. Consequently, the increased deployment of IT allows for greater fluidity of information flows in and across the organizational and institutional setting. This places heavy pressure on decision-makers to communicate and to share information to a much larger extent than in the past.

A key argument in this chapter is that IT is more than a simple tool, enabler, or facilitator of change. The argument is not, however, a technologically determinist one. Orthodox technological determinists believe that the use of technologies determines the outcomes - that the technologies are the *cause* of the change. We argue here that the use of IT tends to lead to certain kinds of changes in organizations and that the impact of the IT predispositions will become increasingly important as IT usage in public organizations becomes more pervasive. It must be recognized, however, that a variety of other factors within public organizations, including political, structural, managerial and cultural factors, shape significantly the use and impact of IT.

Each of the three propositions set out above is examined in detail below. While the predispositions often reinforce one another in their influence on organizational and managerial change, for the sake of analytical clarity, each predisposition is discussed as a separate proposition.

Seeking Skills and Finding Funds

The use of IT predisposes public organizations to acquire specialized skills and substantial capital investment. Meeting the financial and human resource requirements arising from the adoption of IT has become an increasingly significant challenge for governments. Departments and agencies have to compete within their own government for public funds to support IT investment and with the private sector for technologically skilled employees.

The use of IT in the federal public service progressed from a relatively modest emphasis on the automation of clerical functions in the mid- to late 1980s to a major effort in the 1990s to reform service delivery and internal administration. This greatly increased enthusiasm for IT was based on growing recognition that IT could transform radically the way government conducts its business. Between 1986 and 1992, the federal government's investment in IT grew at an average of over eight per cent annually to a total of more than $2 billion. During this same period, $11 billion was spent obtaining IT goods and services from the private sector. By 1993, the IT community in the federal government consisted of about 20,000 employees, with total salaries of about $1 billion. These employees were responsible for installing, managing and maintaining more than 200,000 workstations, over 500 mainframe and mini-frame computers, many different computer-based financial and administrative systems, and hundreds of IT projects.

The robust spending on IT goods in the federal government during the 1980s and early 1990s reflected a shift from the use of centralized computer systems to personal desktops and, to a lesser extent, the automation of service-delivery functions. The office automation trend gradually began to extend to service-delivery points. For example, in 1992 the Department of Employment and Immigration created centrally located automated job information centres (JIC), enabling individuals to search for job vacancies on-line, instead of having to rely on job posting boards at employment centres. At the end of 1993, IT was being applied across 1,200 different federal programs and services. If at that time all federal IT activities had been amalgamated into a single department, it would have had the fourth largest budget in the government behind Finance, Human Resources Development and National Defence.

Since that time, federal government spending on IT goods and services has increased considerably. According to a "conservative" estimate by the auditor general, annual IT expenditures rose almost fourteen per cent - from about $3 billion in 1995-96 to $3.4 billion in 1998-99 - and spending on IT as a proportion of the federal government's overall budgetary expenditures rose from 8.6 per cent to 10.1 per cent. As early as 2000, one estimate put the true annual IT expenditure in the federal government at approximately $5 billion. Federal investments in IT have ranged from the purchase of small volumes of low-cost goods to large purchases for extensive multiyear projects that often involve services from different levels of government and suppliers from the non-governmental sector of the economy.

The increased use of IT has been accompanied by a steadily growing infrastructure dedicated to developing and managing IT tools, frameworks and policies and acquiring a large number of technologically skilled employees. A central feature of the expanded infrastructure is the office of the chief information officer (CIO). The precursor of the cio was the office of the chief informatics officer, which was created within the Treasury Board Secretariat in 1993, strengthened in 1994, and then replaced by the office of the cio in 1996. The cio, who holds the rank of assistant deputy minister and heads a separate branch of the TBS, is responsible for "providing leadership, coordination and broad direction in the use of IT; facilitating enterprise-wide solutions to horizontal IT issues; and serving as technology strategist and expert advisor to Treasury Board Ministers and senior officials across government."

These developments reflect the growing need for a pool of skills and funds to develop and manage IT. For example, Government On-Line (GOL) -originally an initiative and a division within the Chief Information Officer Branch (CIOB) - was given responsibility for allocating $880 million to government departments over six years (2000-05) for the three major purposes of 1) ensuring that the most frequently used services are made available online; 2) creating web sites that provide one-stop access to government services; and 3) developing the policies and infrastructure needed to protect the safety and security of the user.

The increased use of IT requires major investment not only in financial resources but in human resources as well. A competent IT workforce is critical to government's efforts to develop and implement IT policies successfully. The role of IT professionals is evolving from a support service to a strategic resource, as IT professionals are increasingly expected to complement their skills in implementing IT policies with managerial, communication and problem-solving skills. In addition, IT executives (i.e., chief information officers and IT heads within departments) are becoming an important component of senior management.

Governments are in keen competition with the private sector for knowledge workers in general and IT executives and professionals in particular. In recent years, the dot.com bust has made more IT workers available to government, but the task of recruiting and retaining them, especially at senior levels, remains a difficult one requiring special incentives, including higher salaries. The CIOB developed a human resources framework for information management and information technology to help train and retain IT professionals and executives. One of the carrots that government offers current and prospective IT workers is the unparalleled "challenge and scope of IM/IT work in government.... Government is tackling IT projects on a range and scale that is difficult to match anywhere, and in many areas it is charting new territory."

The scarcity of human and financial resources in the IT sphere has obliged governments to rely substantially on the private sector for specialized skills and capital investment through the use of contracting-out and public-private partnerships. John Langford and Yvonne Harrison note the widely accepted view that *"go-it-alone* ICT/e-government strategies are costly and have not achieved the desired levels of performance or transformation desired by governments. Governments are seen to lack IT competence, to have difficulty 'keeping up' on their own with the changing new economy, and to lack incentives to innovate." By contracting-out IT services to the private sector, governments can reduce their need to attract, train and retain well-paid IT workers. Similarly, public-private partnerships can help governments obtain expertise and capital funds for innovative IT projects that would otherwise be poorly resourced or never pursued. The scarcity of resources can stifle innovation by discouraging public servants from undertaking imaginative, but expensive, IT initiatives. It must be recognized, however, that the use of public-private partnerships has become increasingly controversial and that some of these partnerships, especially for large IT initiatives in government, have been costly failures.

Many federal departments and agencies have turned to the private sector for help in the design and delivery of their on-line services. Michelle d'Auray, formerly Canada's chief information officer, has noted that "[a] body of practical knowledge and experience exists in the private sector as does the technological and information expertise.... Furthermore, the private sector has access to financing to augment what government is able to provide by way of new investment dollars." For large IT projects, governments in Canada and elsewhere are utilizing variations on the common service procurement approach. This approach, among other things, authorizes departments to find private-sector partners to supply skills and funding and to share in the projects' risks and benefits.

A notable IT public-private partnership is *Atlantic Canada On-line* - an electronic information service that enables clients to receive a wide array of services. The four Atlantic provinces signed a seven-year partnership agreement with Unisys Canada, which was given a licence to access provincial government databases and provide service to clients (e.g., personal property registration). The governments retain ownership of the information and control of public policy issues, while Unisys is responsible for building and operating the service. Unisys is investing the full $10 million needed to develop, implement and manage the initiative in return for a share of the revenues generated from transaction fees paid by clients.

Another provincial example is the *ServiceOntario* self-serve kiosks, which involve a partnership agreement between IBM Canada and four Ontario ministries to provide information services (e.g., licence-plate renewals) to citizens through a network of electronic kiosks. The project was developed without any provincial funding. All the technology and other capital assets needed to build the network were pioneered and provided by IBM. Under the agreement, IBM has an ongoing responsibility for the maintenance of the kiosks. In return, it is compensated for each kiosk transaction, with the amount of the commission diminishing over the life of the contract.

Since departments and agencies have to fund IT projects in large part from their own budgets, the need to secure alternative sources of capital will continue to drive them towards out-sourcing and partnerships. The development of large-scale innovative projects will also reinforce the need for private-sector participation because host departments and agencies will require cutting-edge technology and skills that can only be provided by technology firms. As we have seen with arrangements such as *Service-Ontario,* governments will also solicit assistance from the private sector in operating and maintaining the technology over the life of the arrangement.

An excellent emerging case of the influence of increased IT usage on the need to acquire skills and investment is the creation of Canada Health

Info-way, a non-profit corporation that, among other objectives, aims to enable fifty per cent of Canadians to have electronic access to their patient record by 2009. The corporation has recognized the need to train physicians so that they can manage information and IT. And the federal government has already committed over $1 billion to the corporation that will leverage its investment through partnerships with both the public and private sectors. The potential commitment of governments to IT in the health sector can also be illustrated by the US$11 billion IT project that is under way in England's National Health Service. This project aims to create a national system of patient records by 2010. *The Economist* notes that the government's poor record for large IT initiatives has led to reliance on private-sector suppliers to provide input and access to the national health records database.

Improving Efficiency

The use of IT predisposes public organizations to improve efficiency by streamlining the management of information. Computing technology has become a major tool for effective information management, defined as "the means by which an organization efficiently plans, collects, organizes, uses, controls, disseminates and disposes of its information, and through which it ensures that the value of that information is identified and exploited to the fullest extent." Applying IT to the management of information can bring substantial efficiency gains. Jane Fountain claims that "[mjovement from paper-based to web-based processing of documents and payments typically generates administrative cost savings of roughly 50 percent - more for highly complex transactions," plus "additional savings of money, time travel, and efforts to citizens and intermediate institutions."

Efficiency gains became increasingly evident over the period from the development of mainframe computing to the current use of advanced computing technology. In the 1950s and 1960s, the adoption of mainframe computing to carry out clerical and administrative functions led to increased productivity while reducing jobs and expenditures. Computer-based systems of information handling can perform many government functions more quickly and more inexpensively while using fewer human resources. For example, sending a forty-two-page document via e-mail from Ottawa to Tokyo is 720 times faster and 355 times cheaper than overnight delivery.

While IT is usually expensive to acquire, its ability to produce and reproduce information goods inexpensively can bring a sizable return on investment. The production of information goods involves high fixed costs but low marginal costs. The cost of information processing has decreased enormously over the past twenty years and is expected to continue to decline. Current estimates show that cost-performance ratios are declining at a rate of twenty to thirty per cent a year. The savings resulting from iT-enabled improvements in IM can help to offset the labour and capital costs arising from the use of IT that were discussed in the previous section.

If the federal government follows through on its commitment to manage information as a public trust and to do this largely through the use of ICTS, significant cost-savings and a related reduction in employment levels should result. The more that IT is strategically managed on an enterprise-wide basis, the more likely it is that cost-savings will be achieved. However, for many IT projects, hard evidence on efficiency gains and employment reductions resulting from increased use of IT is unavailable. Much more common than solid evaluation studies are claims and projections, especially by politicians, of substantial savings and reductions. For example, to effect savings by streamlining government processes and redistributing staff, federal politicians and senior public servants have called for the creation of "shared services organizations" (SSOS), along the lines of those already established in Alberta, British Columbia and Ontario, to allow information resources to be centrally managed across the entire government. This shared service approach would bring together the internal services usually provided separately by each department, thereby allowing departments to concentrate on program delivery rather than on administrative requirements. If all federal departments adopted the sso model, the projected savings would amount to more than $2.5 billion over ten years. The Department of Public Works and Government Services would be responsible for implementing the sso model, but even the CEO of the department's Information Technology Services Branch has noted that implementing the model would involve "a fairly major transformational change in government, in terms of how people do business. It [would] mean a fairly major cultural change in terms of trusting someone or some organization or some entity with things that are particularly important to you - to doing your business."

The scale of claimed efficiency benefits can differ considerably. For example, the government of British Columbia announced efficiency gains arising from a reduction of thirty-five clerical positions because of automation in its Corporate Registry. On a much grander scale, Britain's chancellor of the exchequer, Gordon Brown, announced in 2004 that the government's investment of £6 billion in new IT would permit it to cut the number of civil servants by 84,000 by 2008, while at the same time delivering more efficient public services. The OECD has warned that its member states "have difficulty *monitoring and evaluating* e-government (including costs, benefits, and level of demand), yet increased support for e-government projects will depend on these measures."

Sharing and Disseminating Information

The use of IT predisposes public organizations to share and disseminate information. The concept of "informating," discussed earlier, indicates that IT usage not only promotes efficiencies by improving productive and administrative processes but also generates new information that can be widely shared or integrated. For example, information held in databanks in different parts of a government can be cross-matched or integrated to produce valuable new information for policy-making and program delivery - and, subject to privacy considerations, this new information can in turn be shared across government.

Most public-sector activity relies heavily on the dissemination of information, and most of the processing and communication of this information is accomplished by digital means. Computers, networks, the Internet, fax machines, e-mail and similar ITS have had an enormous impact on the performance of public organizations. Information technology has dramatically reduced the historical barriers to disseminating and sharing information. By spreading information with remarkable speed, IT is dramatically reducing time and distance barriers. Stand-alone computers, as the name implies, are largely self-contained, with no real communications component. By comparison, computers that are networked act as a system that enables the information to be widely disseminated among users.

The spread of IT resources in the public sector has been associated with a sharp increase in the level of intra-organizational and inter-organizational communication. The high rate of Internet penetration in government has been accompanied by public servants' extensive and routine use of e-mail. Each day about six million e-mail messages are exchanged within the federal government. Information can also be spread quickly through a government's *Intranet,* which provides employees with a direct and streamlined mode of interaction with other employees across the government. In addition,

Intranet-enabled training can reach a large number of staff in a short amount of time and the Intranet is useful for informing staff of the latest news or issues, serving as a platform for the publication of internal documents and even facilitating teleconferences. Computing technology has also enhanced management information systems that now provide decision-makers with much more information and on a real-time basis. Information technology enables public servants to interact more easily and more frequently and thereby encourages a high degree of information-sharing.

The use of IT also encourages the dissemination of information across organizations - and across governments and sectors as well. The availability of information made possible by computing technology and the development of government-wide IT infrastructure promotes the sharing of information across departmental boundaries and a consequent reduction in overlap and duplication of internal administrative and service-delivery functions. For example, the electronic directory service launched by the federal government in 1996 to provide government-wide directory information on public servants (e.g., name, telephone number) has produced substantial cost-savings through paper reduction.

The increased sharing and diffusion of information across governments and sectors can be effectively demonstrated by reference to the remarkable growth of integrated service delivery (ISD). These arrangements bring together government services so that citizens can meet their wants and needs seamlessly. Integrated delivery provides single-window service (one-stop access to services), largely through collaboration across governmental and sectoral boundaries. These collaborative arrangements are driven not only by iT-enabled sharing and dissemination of information but also by governments' needs, discussed earlier, to obtain specialized skills and substantial capital investment through public-private partnerships. Collaboration can involve a large number and variety of participants. For example, the Canadian Consumer Information Gateway joins up more than 400 partners, drawn from departments and agencies in the federal, provincial and territorial governments and from non-governmental organizations. Partner organizations are able to maintain their own information on the main portal by remotely disseminating their collections to the central databases. This ensures the currency of information and provides more functional value than if partners were simply to post links from their existing home pages.

The use of IT has enabled ISD arrangements to take a wide variety of organizational forms, thereby supporting the view of those who contend that informatization adjusts or changes the organizational structure in which IT is introduced. These new arrangements include ISD units within departments (e.g., as in British Columbia, Ontario); corporate service utilities that take the form of a public corporation, with a board of directors reporting to a departmental minister (e.g., Service New Brunswick); NGO service utilities, which involve a non-profit organization with governmental and private sector partners (e.g., Victoria Connects); and organizations providing service delivery through integration (e.g., the Canada/Manitoba Business Service Centre, which has thirty-six federal, twelve provincial and eight nongovernmental partners).

Barriers

While the use of IT predisposes public organizations to share and disseminate information, there are, as suggested earlier, substantial obstacles to realizing IT's full potential. Moreover, the extent and impact of the obstacles will differ from one government, one government department, and one policy sphere to another. The scope and nature of these obstacles are illustrated here by reference to the challenge of using IT to promote ISD across departments and governments.

Among the *political* obstacles to ISD in Canada's federal and provincial governments is the fact that the individual responsibility of ministers for the conduct of their departments encourages them to focus on the vertical dimension of government. The horizontal perspective that is needed to foster ISD through the Internet channel is relatively neglected. *Structural* obstacles include the related problem of "departmentalism." The traditional "silo" design of departments runs strongly counter to the current emphasis on horizontal government: "Governance structures favour departmental projects, not interdepartmental partnerships."

Legal obstacles to ISD include legislative and regulatory barriers that require political consent for their removal or modification. Some of these obstacles are more evident than others. It is clear, for instance, that privacy acts restrict the sharing of some kinds of data between government and business partners. Less obvious is a situation like that involving employees of Canada Customs and Revenue Agency (CCRA), who have immunity in respect to giving tax advice - an immunity that does not extend to ISD partners who may act on the agency's behalf.

A major *operational* obstacle is the lack of inter-operability, in the sense of differences across departments and governments in laws, policies, rules, regulations and standards. Multiple accountabilities have to be sorted out; different human resource policies, including salary levels, have to be accepted or accommodated; different unions representing staff from different jurisdictions have to be satisfied; and financial rules have to harmonized.

Among the *managerial* obstacles to effective ISD is the lack of resources arising from the reluctance or inability of governments to fund projects on a horizontal basis. Integrated service-delivery projects are often multiyear initiatives that require substantial up-front investment and take considerable time before savings can be harvested. Changing a department's or a government's budgetary system to permit dedicated funding of horizontal projects can be extremely difficult.

Finally, *cultural* obstacles can inhibit the blending of organizational cultures required by many ISD arrangements that extend across departments, governments and sectors. Long experience working in silos under strict accountability requirements creates a culture of tunnel vision rather than the peripheral vision needed for horizontal government. This tunnel vision is often accompanied by turf tension, as individuals and organizations strive to protect established mandates and processes, in part by restricting the sharing of information.

A major impediment to the sharing and dissemination of information that cuts across several of these categories of barriers is concern about the infringement of individual privacy. The many aspects of this large issue cannot be examined here, but special note should be taken of the continuing tension between protecting citizens' privacy and using IT to facilitate transparent and efficient government. E-government leaders are devoting substantial resources to reducing this tension; however, a substantial number of citizens are still reluctant to provide information and conduct transactions on-line that may jeopardize their privacy.

Implications for Middle Management

This third predisposition, involving the sharing and dissemination of information, is closely related to the second predisposition, in that it promotes the streamlining of IM and can, thereby, foster organizational efficiencies. In particular, the two predispositions can reinforce one another to foster a reduction in the ranks of middle management. Extending the lines of communication to allow senior management to send and receive information to and from lower levels of the hierarchy can reduce the need for middle management. A de-layering of the middle levels of the hierarchy, which promotes cost-savings, can flatten the organizational structure. However, the extent to which IT contributes to a reduction in middle management is a matter of debate. Alain Pinsonneault and Kenneth Kraemer reviewed the empirical work on this subject to 1990 and conducted their own empirical study in 1997. They concluded that if both computing and organizational decisions are centralized, top managers tend to use IT to reduce the number of middle managers, but if these decisions are decentralized, middle managers use IT to increase their numbers. Sandford Borins and David Wolf assert that since 1990 the use of IT has resulted in more substantial reductions in the number of middle managers than before that time.

With particular reference to the public sector, John Taylor argues that "[t]he capability to transmit information directly from operational to strategic levels of the organization means that fewer 'booster managers' are required in the middle tiers of the organization than before information networks were installed." And Borins and Wolf note that in a period of government downsizing "we could expect that [pressure to reduce the number of staff] would result in a reduction of the ranks of middle management in all departments, but that in the more operational departments, where information roles can be replaced by IT, this reduction would be the greatest." In support of this hypothesis, they cite the findings of the 1996 U.S. National Performance Review that surveyed changes in the number of supervisors and management positions across government. The survey found that departments with the smallest percentage of reductions in middle managers were "central agencies" (e.g., Justice, the Treasury), whereas those with the largest percentage of reductions were operational departments (e.g., Agriculture, Defense). Thus, the increased adoption of IT can be expected to contribute to flatter organizational structures, especially in operating departments, by encouraging the de-layering of the middle-management ranks.

The changes in the structures and processes of public organizations that are influenced by the three IT predispositions are closely related to the major features of informatizarion set out at the beginning of this chapter. Especially notable is the impact of the predispositions in respect to

1. *the adjustment or change of the organizational structure in which information technology is introduced* (e.g., integrated structures for service delivery, the reduction of middle-management positions);
2. *the development of information policies as a differentiated area of decisionmaking in the organization* (e.g., the creation of chief information offices in many governments, new policies on the management of government information); and
3. *the introduction of specific expertise in the field of information technology through functionaries or consultants with specific tasks in this field* (e.g., reliance on private-sector skills, the use of public-private partnerships).

The three IT predispositions also influence the movement of public organizations between the bureaucratic and post-bureaucratic models of public organization. Especially notable is their impact on the movement

1. *from independent action to collective action.* The need for special skills and capital investment and the pressure to share and disseminate information reinforce the movement of departments and agencies towards collaborative arrangements, including public-private partnerships and various forms of ISD.
2. *from the departmental form of organization to non-departmental forms.* Again, the need for expertise and investment combined with the capacity to share and disseminate information is prompting the creation of non-departmental forms of organization, particularly ISD arrangements such as service utilities.
3. *from hierarchy and central control to decentralization of authority and control.* The displacement of middle managers helps to flatten the hierarchy, and the sharing and diffusion of information fosters decentralization of authority and control through collaborative arrangements between departments, governments and sectors. However, IT can also have centralizing effects in at least two significant ways that move public organizations towards the traditional bureaucratic model of central control and coordination. First, as Fountain argues, compared to the traditional Weberian emphasis on rules and procedures, contemporary information-based organizations can be just as strongly, or even more strongly, rule-based and rationalized: "[R]ules embedded within information systems become less visible and

seemingly less constraining to bureaucratic discretion. Embedded rules will increasingly replace overt supervisory control." Moreover, Ignace Snellen notes that the increased use of "text-blocks" - prefabricated justifications of decisions - is tending to narrow the discretion of street-level bureaucrats. The use of textblocks involves public servants entering the facts of the case into the information system, which then suggests not only the decision but also the reasons for it.

The use of IT has centralizing effects in a second respect. The pressure to coordinate or integrate service policy and delivery, and thus to ensure the inter-operability of computer systems, adequate IT expertise and compatible rules and standards, argues for centralization of IT authority and resources, both within departments and across government. This pressure runs against the current trend towards devolution of authority and control from central agencies to departments and within departments themselves.

We have examined three propositions concerning the implications of IT for public organizations in Canada. This examination supports the argument that the process of informatization is having a substantial impact on these organizations, in that they "are becoming strategically and centrally dependent upon the changing flows of informational resources which are made possible by powerful combinations of information and communication technologies." We have argued that the use of IT by public organizations predisposes them to change in particular ways. However, account has been taken of the "dark side" of these changes in the form of such factors as inadequate evaluation of claimed efficiency gains, undue reliance on the private sector for specialized skills and capital investment, and major privacy issues. Political, cultural and other obstacles to public-sector reforms involving IT have also been identified and illustrated.

Dealing with constraints on the advance of informatization will require both short-term reforms and longer-term reforms. The OECD, in its checklist for successful e-government, invites e-government leaders to ask whether there has been a review of barriers to e-government implementation and whether e-government has been integrated into broader policy and service-delivery goals, into public management reform goals and processes, and into broader information society activity. Canada's politicians, public servants and academic scholars need to collaborate in answering these questions so that IT can be effectively integrated into the practice and study of public administration.

REFERENCES

Paul H.A. Frissen, (1992). "Informatization in public administration: An introduction," *International Review of Administrative Sciences* 58, no. 3, p. 308.

Bellamy and Taylor, "Informatization and new public management," *Public Policy and Administration,* p. 29.

Shoshana Zuboff, (1984), *In the Age of the Smart Machine: The Future of Work and Power* New York: Basic Books, p. 9.

Jane E. Fountain, (2001), *Building the Virtual State: Information Technology and Institutional Change,* Washington, D.C.: Brookings Institution Press, p. 5.

Borins and Wolf, "Realizing the Potential of Public Sector Information Technology," in [Thomas, d'Ombrain, Borins and Wolf], *Change, Governance and Public Management,* p. 182.

Bellamy and Taylor, "Informatization and new public management," *Public Policy and Administration,* p. 29.

2

Information Technology and Public Sphere

Information technology and globalization have each been the objects of enormous hope and considerable disappointment. So too is their combination in the notion of an international public sphere supported by the Internet and other communications media. This is basic to the dream of international civil society that has flourished since the early 1990s as the collapse of communism and opening of capitalist markets. And indeed, such an international public sphere clearly already exists. Equally clearly, however, it has not yet provided the basis for cosmopolitan democracy its advocates have hoped.

Globalization

Information technology has of course been an important basis for globalization. This did not start with the Internet, despite its hype. From the sextant to the map to the invention of calculus, the development of early modern navigation was centrally a matter of information technology—and the ships circulated information as well as goods and helped to link the globe in networks of exchange and control. Telegraphs, radio, and television; trains, cars, and airplanes all helped to establish links across space and shorten the time-lags to action at a distance. The development of national and international highway infrastructures paved the way to flows of information long before the "information superhighway". Colonial empires were pioneers in information technology from at least the 16th century. They developed bureaucracies, accounting systems, file-management hardware and software, and both open and secret networks of communication. They also remind us that new forms of war figure centrally in the story of modern globalization—with new weaponry, certainly, but also military reliance on ever-more complex systems of control and coordination managed by technologies that evolved quickly from carrier pigeons to coded telegraph messages to satellite transmissions. As the last suggests, too, globalization has all along included an element of surveillance, which has always been a matter of data management and analysis as much as observation, and which has benefited from technologies to improve each.

Old though some of the trends are, the last twenty years' innovations in computer and communications technologies have also been powerful. Modern financial markets are inconceivable without computer mediated communication—and action. Not only are there nearly instant flows of money through global networks, ever more complex forms of derivatives are driven by algorithms their "owners" never know and trading is done indirectly by software "robots" programmed by their human "supervisors". Satellite transmissions and electronic filing of stories speeds up the flow of information through broadcast and print media. New Internet media complement older forms of circulation of public (and sometimes not so public) information.

Migrants maintain global diasporic cultures and translocal communities partly by email and websites. The Internet is an important support also for global social movements and for local social movements seeking global attention or support. It is important, though, not to imagine that with electronic help popular forces can easily get the jump on more centralized or elite powers—whether economic or political. For all the energy and innovation of global activism challenging neoliberalism or inequality, at least as much and a good deal more resources are deployed in support of global capitalism.

NGOs focus not just on human rights or the environment but on accountancy standards, corporate advocacy, and arbitration between firms. And information technology is employed intensively in organizing global production processes as well as global markets. It makes possible not only just-in-time shipping but centralized control of manufacturing facilities that are physically dispersed throughout the world. What distinguishes the last thirty years or so, as Manuel Castells puts it, is the creation of "an economy with the capacity to work as a unit in real time on a planetary scale." This is a "global" rather than merely a "world" economy. It is also the basis for

Castells' metaphor of the "network society", which refers not simply to the fact of global connections but also the adaptive advantage of flexible, network-based forms of organization in relation to dynamic, global capitalism.

Global interconnections still include war, of course, which itself is transformed by the rise of flexible, network organizations. This is covered by nearly instantaneous press reports and often carefully managed visual images that make it a broadcast spectacle. In three senses it is appropriate to speak of information technology contributing to the emergence of "virtual war". First, there is the conversion of warfare into a media spectacle woven into the same fabric as Hollywood war movies. Second, there is the blurred line between war games and automated strategic response as Pentagon planners rely on computer models. Third, there is the waging of air wars that keep casualties all but invisible—for those in the planes or back at the missile bases. In this second sense, war may be "virtual" for the United States with its enormous technological superiority and military strategy premised on the notion that American casualties cannot be tolerated even while it is all too "real" for those suffering on the ground. The Persian Gulf War inaugurated this new asymmetrical virtuality; it was honed in Kosovo and deployed effectively in Afghanistan. But virtual war is not the whole story. New forms of war increasingly involve nonstate actors—famously, the terrorist network Al Qaeda.

These sometimes rely heavily on the Internet to maintain organization and plan action while dispersed. They also use both broadcast and computer-mediated communications to disseminate their message and recruit new adherents. While "virtual war" sanitizes killing by technological mediation and keeps bloody violence at a distance, terrorists rely on exactly the opposite use of technology. Violence is not simply a way to accomplish military objectives for terrorist networks, it is one of their objectives in itself. Footage of the World Trade Center attacks offered Al Qaeda and similar groups one of the world's greatest recruiting videos. And part of the message was that technology could be a double-edged sword, one that those who considered themselves oppressed could master and use against its creators.

So far, cyberterrorism has been minimal and largely symbolic—as for example Indians and Pakistanis flooded several sites in each other's countries, demonstrating not only enmity but also the capacity to penetrate security systems. This capacity is relatively widespread, but still unevenly so. Distribution of skill may be one reason for the low prominence of cyberterrorism. While few if any discontented populations lack members with the skills to hack into websites, organizing a major destabilizing of social systems by means of attacking a technological system is more complex than merely hacking in. More important, perhaps, is that while cyberterrorism has begun to figure in science fiction it has not yet lent itself to spectacular representation. A key feature of the World Trade Center attacks is that they both imitated film and made good film. They followed a template for a spectacular public devastation that was more widely available and thus easier to publicly conceptualize and communicate about than cyberterrorism. While an attack on a computer system might be a more effective way to bring parts of the global financial system to crisis than an attack on buildings, it wouldn't be equally capable of immediate, clearly interpretable global representation. It would in important senses be a private attack on the public world, not a public attack.

It is also important to note the extent to which global discussions of IT and its regulation have come to be embedded in security concerns. Especially since September 11, security (and military, intelligence, and police cooperation) have moved to the forefront of the agenda among those organizing interstate connections. This is likely to have long term implications for the development of IT use, which is contingent in important ways on the legal-regulatory infrastructure created. This involves issues about security and surveillance as well as property rights which are seldom openly debated in the public sphere.

Alongside war, and partly spread by war, disease also figures prominently in the "dark side" of globalization. This is not just a matter of possible bioterrorism. More people die daily of AIDS than died in the World Trade Center attacks. And if the spread of deadly viruses from the less well developed (and thus less healthy and less vaccinated) world is a source of anxiety in rich countries, it is nonetheless the poor who suffer disproportionately from nearly all diseases. And disparities in access to treatment become one of the most dramatic faces of global inequality. Here too information technology matters, though what we see most is the weakness of the global public sphere in spurring effective action. We see also the power of Western and especially American media. These portrayed AIDS largely as a gay disease, thus adding a stigma to the difficulties of fighting it in the rest of the world—where homosexual contact is a minor vector compares to heterosexual transmission, drug use, and unsafe blood supplies—and as more or less 'handled' by drugs available only at a cost prohibitive to much of the world.

In this context, it is worth noting that the most influential of electronic media is not yet the Internet or any form of computer mediated communication but still television. Even its reach is not altogether global, but it comes close. Access to and use of the Internet are still much more uneven, and although usage of the Internet grows it is less basic to shaping the broad "information background" to all public discourse.

The different technologies are mutually intertwined, of course, not only with Internet sites run by television broadcasters but with news reports gathered by Internet. Just as television often alerts people to stories they will read in newspapers, so an email message can tell someone to turn on the TV or a TV show can send someone to the web for more detail. Aside from inequalities of access, there may be even more telling inequalities in production of content for various media. Though access to the Internet is uneven, it does allow for point-to-point communication, dispersed access to common information resources, and relatively inexpensive posting of new content. Broadcast media remain almost completely one-directional, with broadcasters determining the form in which content appears and with the costs of competitive entry high.

On the other hand, while more and more information is available on the Internet there is a wealth of globally significant knowledge still accessible only in the libraries and computer data centers of rich countries. The results of most social science research on the world's less developed countries, for example, are accessible only in the knowledge centers of the rich countries. Moreover, commercial sites tend to drown out nonprofit ones oriented to the public good and search engines not only miss much of the web but generally relegate small scale sites to their back pages.

Global news media are heavily controlled by a small number of Western corporations and public broadcasters from a few Western governments. This is mitigated somewhat by the prevalence of national broadcasters, though without their own substantial news-gathering operations the global content these can provide is limited to what they can acquire from the major global providers. One of the interesting sidebars to the post-September 11 "war on terrorism" was the emergence of al Jezeera as an important global media player. This Arabic network not only broadcast throughout the Arabic-speaking world it became a key source of content for the major American networks and European broadcasters. Much Western commentary initially treated al Jezeera with suspicion, viewing it as "an Arab voice" while the Western media were treated as neutral or universal. That al Jezeera became a conduit for the messages of Osama bin Laden fueled this perception, but it may be more apt to see al Jezeera as part of the emergence of a transnational Arabic-language public sphere.

All the electronic media, like books, newspapers, and the International Postal Union before them play a crucial role in extending communication beyond local, face-to-face contexts. They thus underwrite globalization. Again, though, there is a tendency in some of the speculative literature of early enthusiasts to imagine that the new media turn the tables on traditional inequality of access to information and effectiveness of communication more than the do. It is certainly true that international activism of indigenous peoples, environmentalists, and opponents of the WTO has been organized in new ways and with greater efficacy because of the Internet. It is equally certainly not the case that such use yet rivals the efficacy of information control by transnational corporations. Both the corporate control of much public content provision and the corporate use of information technology to manage internal production and private financial transactions so far considerably outstrip insurgent and activist uses of the new technologies. This doesn't make the latter unimportant, but it should encourage a certain realism. As with other technologies, the ability to use IT varies not just with the potential of the technology but with the resources different users can invest.

The very ubiquity of IT has another curious effect on the global public sphere. It is part of a construction of globalization as an inevitable result of technological progress. From different national vantage points, the question is commonly posed not as whether to join in this globalization but how to adapt to it. Challenges to the dominant Western— indeed American—neoliberal, capitalist forms appear simply as backward-looking traditionalism. This was perhaps especially true during the economic boom of the late 1990s; how visions of the future and struggle over capitalism will fare in less soaring economic times is unclear. But IT continues to plan a double role—as the visible face of high technology and as part of the technical underpinning of a greater awareness of global trends.

In many countries around the world, the response is to try to adopt new technology as rapidly as possible, while simultaneously trying to protect "traditional" culture. Contemporary Indian politicians thus project their country as a potential information technology superpower at the same time that they encourage a renewal of studying the Vedas and a more or less fundamentalist embrace of Hinduism. It is precisely the most culturally conservative party to lead the country in its modern history that announces the most high tech economic plans—as part of its "Vision 2020" proposals. The very phrase "Vision 2020" is not uniquely Indian, however, having been employed in Malaysia and elsewhere. The common rhetoric pairs technology-grounded progress with protection or renewal of cultural traditions. The formula is old, being something of an update of the 19th century Chinese response to the West: "Western learning for material progress, Eastern learning for spiritual values" (a phrase commonly summarized by the syllables *ti-yong*). But as the Qing emperors learned, it was hard to successors import railways and telegraphs without bringing Western values along. China's communist leaders worried about the same issue in the 1980s, and both their suc and some contributors to China's popular websites worry about the same thing today. Even Canadian politicians echo the same theme, showing that anxiety about local identity has as much to do with American—that is, United States—power as with a clash between the West and the rest. As government officials told *Financial Times* reporters, "the country wants to become a lean global competitor while maintaining traditional local values".

Politicians presenting globalization as an irresistible force commonly embrace neoliberal policies as the necessary response. Global economic competition demands, citizens are told, that public sector jobs be cut, taxes lowered, state-owned resources be privatized. The same information technology that potentially makes possible a more vigorous public sphere is cited as an example of global pressures before which public opinion should bow down; it is used to squelch dissent. Ironically perhaps, the IT industry itself often backs this neoliberal rhetoric, focusing on the image of freedom and seeing states and bureaucracies only as possible fetters on innovation (rather than also embodiments of social achievements such as public education or health care). The result is to reduce the chances of the socially transformative and publicly engaged use of IT, while encouraging its domination and domestication by commercial interests.

Nonetheless, IT use does escape being harnessed entirely by capitalism. Some of this depends on slipping out of the net of property rights in order to enter the World Wide Web. In India, by some estimates, the majority of people make access to the Internet illegally, from informal sector Internet vendors who establish hidden or black market access to the system's backbone or trunk lines. Without this—and a variety of other shady but creative adaptations around the world—IT access would be even more asymmetrically distributed than it is. But the solution is only partial. It enables email and bulletin boards and access to public information. But it does not speak to the extent to which the provision of content on the Web is itself ever more commercialized.

Both the flow of information and the financial value of Internet communications put IT on the agenda of government and multilateral regulators. Information flows on the Internet raise concerns about security, manipulation and crises of unintended consequences (such as the financial flows that were central to the East Asian fiscal crisis of 1997). So far, there has been relatively little public discourse about the nature of this regulation. Indeed, nongovernmental advocates for the public interest have largely been excluded from the discussion. When the G-7 powers established a working group on Internet policies in 2000, for example, they determined that each government would have one representative and that there would be a representative of each national private business establishment. Only under pressure did they agree to "observers" from less developed countries. Efforts to secure representation for the nonprofit sectors in the different countries failed. This brings up the basic issue of whether there is an effective way for non-business civil society interests to be heard. The open source software movement has made remarkable strides, but advocates for this sort of vision of large-scale cooperation in the creative process, and of a more democratic approach to developing a technological infrastructure are not significantly represented in core discussions of legal and political regulation. National public spheres and the nascent international public sphere alike are sharply challenged by the partnership of corporate powers and national governments, backed up by international agencies serving that partnership and a global culture heavily shaped by neoliberalism.

To the extent that there is a direct response to self-organized public communication, it comes mostly from nation-states. Though businesses are susceptible to campaigns focused on their public image (famously pioneered by the Nestle boycott), they are much less directly attentive to public opinion when it is expressed in non-financial terms. It is the translation of opinion into either consumer purchasing patterns or capital investment patterns or occasionally the willingness of other firms to work cooperatively that matters. Likewise, multilateral organizations like the World Bank, the IMF, and the WTO can be challenged by public outcry, but are governed largely by states and operated along technocratic lines, with expertise conceived mainly within the terms of neoliberal ideology. Accordingly, it is a mistake for democrats to accept too readily the idea that the growth of a global civil society makes states unimportant. States still matter, they still wield considerable power, and this is at least in part a good thing, not least because they are the main locus for democracy.

It is common to speak of globalization as though the fact of worldwide interconnections meant that there was really a single, seamless whole. I have done this to some extent here, but it can be misleading. Globalization looks different from different regions and localities in the world. It is both an opportunity for the EU and a pressure behind European unification. The implications are more ambiguous in much of Africa, where access to the benefits of globalization has been minimal and the legacy of colonialism and other ills have in some cases been exacerbated by it. Instead of wondering whether states will lose their familiar power, for example, many Africans must wonder instead whether states will become strong enough to manage public services and unite countries. Likewise, senders and receivers of migrants must view migration differently. And while unequal access to information technology is an issue everywhere, the inequality is between countries and regions as well as within them. And the term 'access' can be misleading, since even countries that are in a good position to make access to the web may not be in a strong position to become suppliers of global information resources.

To a large extent, the notion of a single, uniform global culture or economy is an illusion encouraged by the way dominant Western media present information about the world. From other vantage points, the divergences are readily visible. Globalization is made up of a variety of different projects that clash with each other in varying degree and combine in different ways in different places. Western dominance is contested. This is not to say that what happens elsewhere is simply survival or response to Western initiated trends; it is rather other the creative development of alternative modernities. This is something of what has been at stake in clashes between Islamists and the West in recent years—not tradition vs. modernity so much as struggle over whether Western trends can control the whole of the modern. Is the freedom to make pornography predominate in the content available on the Web inherent to modernity, or a contingent choice the West has made in constructing modernity—not only in terms of sex but in terms of minimally restricted commercialization. Moreover, Western-dominated globalization is not all of a piece—the WTO and the anti-WTO protests have both been Western-dominated. Neither is there simply a tension between the West and the rest, as though the latter were uniform. Regional and local differences still matter, along with religious, cultural, class, and other differences that cross-cut regions. Even nationalism is commonly a response to globalization, not its other, something old and passing.

Globalization is a heterogeneous process, in short, which produces multiple and overlapping layers of interconnection. It is a matter of enmities and rivalries as well as interdependence. Moreover, the other side of connectivity is exclusion. The more densely certain networks are woven together, the more difficult it is for outliers to find a way in. International civil society has grown as part of globalization, but its enthusiasts must be careful not to overstate its strength. It is sometimes suggested that IT is a great equalizer, but as we have noted IT is used by corporations at least as effectively as protesters (if not a good deal more so). Describing the surge of global protest over capitalist hegemony, unequal development, and the environment some writers seem almost to imply that the protesters outside the WTO meeting in Seattle came close to overpowering the global elite within. In fact, of course, they were able to get a message heard and to create a disruption but not to assume significant organizational power.

IT plays an important role, finally, in making it possible for globalization to combine decentralization and dispersal of activity with concentration of power. As Saskia Sassen notes, much of what official statistics present as global trade among seemingly independent nations is in fact coordinated production within individual global corporations. Carburetors may seem to be sold from Mexico to the US, but this is just one link in Ford's new global assembly line, with computer assisted control systems and financial management as important as computer assisted design or robotics.

Networks and Knowledge

Information technology is powerful, but not all powerful. It is introduced into a world of existing social relations, culture, capitalism, and inequalities. These shape what will be made of it. So does the creativity of engineers, designers, and users. So too, potentially, can the creative work of artists and others for whom the Internet provides a new medium for both production and circulation of work. Indeed, the aesthetic potential of information technology is both important in itself, and potentially part of the contribution of IT to the public sphere. The latter is never simply a matter of rational-critical argument but also of cultural creativity, the reimagining of the nature of social relations can be as important as debate in the life of the public sphere.

Through most of the 20th century, for example, the public sphere joined an aesthetic and a distributive critique of capitalism, a bohemian discourse with the socialist, hopes for a more beautiful world with hopes for a more egalitarian world. To a considerable extent this combination came unstuck in the 1990s. Egalitarian ideals were tarred with the brush of defunct Soviet-style bureaucratization and notions that government regulation was excessive even in the West.

Conversely, capitalism appeared often in the aesthetically appealing guise of the high-tech corporate campus rather than the polluting factory. For a time, at least, its paradigmatic occupation became website designer and not only its economic but its aesthetic potentials seemed unlimited. There was more artistic freedom offered by the dot coms than by government (and even universities looked awfully stodgy). How much of the changed attitude will survive the dot com crash is unclear, but the example nonetheless illustrates the ways in which the revision of cultural images is important to the construction of public opinion. And of course, it was not an accident. The hybridization of Madison Avenue and Bellevue, Washington was central to the dot com era, and focused precisely on the revision of public opinion by aesthetic means.

Attempts by states and others to regulate the Internet and similar technologies may reflect purely bureaucratic concerns or may respond to such public discourse. A significant gap in this process, so far, is that information technology has been the object of a great deal of speculation—often utopian but sometimes dystopian—and not much serious social research. Such research will advance best if there are more researchers with serious knowledge of both information technology and social science. Such "cross-training" is not encouraged in contemporary universities, but is not impossible. In any case, for the public to make informed choices about the development and deployment of information technology—and perhaps about remedying inequalities in access or impacts—knowledge is crucial.

Perhaps ironically, at the same time that more and better knowledge is needed about the social changes in which information technology plays a central role, the traditional institutions of knowledge production, storage, and distribution—like universities—are undergoing an unannounced structural transformation partly based on IT. This can serve as an example for issues in the study of IT. Various commentators have wondered what the implications of new technologies might be. Optimists have pointed to the capacity to reach new students in dispersed settings, to bring currently centralized resources like libraries to learners and researchers in remote places, to put give students more chances to control their own pace and directions of study, and to make institutions more efficient. Pessimists have worried that 'efficiency' will be used only to cut costs not to expand offerings, that personal contact will be reduced, that knowledge work (and curricula) will become more standardized, that this might be linked to more reliance on standardized testing, or that social benefits associated with campus communities will be lost if students are isolated from each other. All of these good and bad possibilities are realistic. Two problems stand between speculation and useful knowledge, however. First, there is little research, especially researching showing not just whether any of these things are happening someplace, but which of them are happening where and in association with what else. Second, we must overcome the tendency to try to figure out the impact of IT in isolation from broader social patterns of which it is part. In this case, the role of IT is currently secondary, but growing, within large-scale structural transformation of higher education that has gone largely unannounced and unexamined—at least within the United States.

Sticking only to the American example (though versions of this process are widespread), several different changes are intertwined and are fundamentally changing the character of higher education. The changes are not evenly distributed, to be sure. Most especially, wealthy institutions have more capacity to shape their own futures—and to maintain attractive features of their current character. Thus it is not likely that Harvard University or Williams College will employ IT simply to cut costs, reduce faculty, or teach students online. They may use their "brand names" to offer profitable distance education programs, but their endowments and social status will enable them to continue to serve a privileged segment of the student population with residential programs that reach far beyond the formal curricula and that include face-to-face relations among students and between them and the faculty. The options open to less well-endowed institutions are different, and so too the pressures shaping them. One reason we know less about the overall transformation of higher education is that researchers tend to be drawn from the upper tier of schools and reflect mainly on what has gone on in the institutions in which they were educated or work. But higher education in the United States is a highly differentiated system in which schools have different missions, student bodies, places in academic labor markets, and resources.

Some of the most noticed innovations in higher education have come with the rise of for-profit schools. These are disproportionately (but not exclusively) distributed in the less selective segments of the higher education market. It is no surprise that DeVry Institute of Technology competes more with community colleges than with Stanford (or that the institutions acquired worldwide by Sylvan International Universities are primarily in the market for first generation university students wanting to study applied fields). It is also significant that providing technical education—much related to IT—has been a mainstay of the new institutions. The older higher education system has been slower to develop needed offerings. But if this is partly simply conservatism, it is partly a question about the nature of education. Colleges and universities have commonly maintained that even technical education was best pursued in the context of broader liberal arts and sciences, that it was important to train citizens as well as technicians, and that students needed to develop critical thinking skills for both roles. In relation to IT, many have pointed out the difference between mastering a specific software package or hardware configuration, both of which are likely to change quickly, and mastering the underlying principles that will enable students and workers to learn about new systems in the future. Absent a much greater social investment, though, this broader vision of higher education is inevitably somewhat elitist; it speaks less well to students worried about the fastest, surest course to a good job.

Traditional colleges and universities are expensive—and are under challenge from "rationalizers"—not only because they are inefficient, but because they bundle together a number of different activities and functions. They house libraries and sometimes publishing companies, counseling services and sports teams, computer centers and hospitals. They provide entertainment, housing, and at least implicitly dating services. One of the basic questions today is which can—or should—maintain this bundling together of disparate activities. Perhaps most basically, the ideal of the university has combined the production and dissemination of knowledge, research and teaching. Some efficiencies are available by unbundling these, and this is happening. Like many of the changes, it is happening within traditional institutions as well as in a redistribution of activities among institutions. Some faculty are able to negotiate privileged reductions in their teaching loads because of their research productivity; others make up the difference. The most extreme dimension of this is the rise in employment of temporary and part-time teachers, usually with low salaries, minimal benefits, and marginal status within the institutions. Though transformations in higher education are often described in terms of "privatization", this shift appears in public institutions as well, and market pressures are felt throughout the system.

Indeed, profit and non-profit orientations are being intermingled. Not-for-profit universities have established for-profit corporations to try to make money from online offerings. University presses long-subsidized because the academic value of their publications are being asked to pay their way as businesses. Charges for some library resources, especially those provided electronically, are becoming more widespread. The primary products of scholarly research—journal articles and books—are increasingly accessible only for a price. The institutions most immune from pressure are in fact not the public universities but the very well-endowed private ones. MIT and Princeton, thus, have made their online offerings free—both seeing them as more valuable for public relations and alumni loyalty than as profit centers, and suggesting that there may not be as much profit in this kind of offering as is sometimes thought. On the other hand, such well-endowed institutions are also operated partly as for-profit businesses, not least when they get into joint ventures based on the inventions and innovations of their faculty.

In this as in other areas where IT is part of a larger change, inequalities are growing but masked. Not only individuals but institutions are unequal when facing technological innovation. Around the world, privatization of higher education is part of the more general process of reducing state support of social goods. This is encouraged by the neoliberal ideology dominant in contemporary globalization. But it means that states—the institutions best placed to defend the gains workers and other popular forces have made in previous struggles—are instead abandoning them. Some of those struggles—such as for low-cost, equitably or perhaps meritocratically distributed public education—have been waged in part through the public sphere, with material pressures complemented by shifts in public opinion. The supporters of neoliberal capitalism have effectively dominanted elite public opinion, including the ways in which the most powerful states perceive the implications of globalization. This means that debate in the broader public sphere is an important test for the capacity of civil society to resist the imperatives of capitalism. The debate influences not only current state actions but the availability of knowledge for future public discourse.

The Internet will always be a supplement to not a replacement for other connections. Those who already have power can make more effective immediate use of the technology—so those with less power have to work harder to make it effective on their side. The Internet makes it easier to find things out, thus, but often harder to find out what we don't know. I don't mean simply that it adds to volume and creates new demands for ability to discriminate and evaluate sources, though this is certainly true. I mean also that it makes new forms of surveillance possible that are inherently hard to for ordinary people to observe and which may nonetheless dramatically enhance the control others have over their lives. Despite the fact that privacy issues have received more research and public attention than almost any other dimension of information technology, surprisingly little public outcry has greeted US government proposals—and actions—to reduce restrictions on such surveillance. People still seem to think mainly in terms of new information being gathered rather than recognizing the immense detail of information already produced as byproducts of computer use but (usually) not linked across data sets. The label privacy is partly misleading, thus, since the issues involve not just "exposure" but control.

Control is also important in another way. It is usually obscure to users who is editorially shaping the content available on seemingly open public websites. China, for example, has a range of extremely lively web-based discussions about issues from international affairs to government corruption. Content that offends the government usually disappears quickly, however, and there is presumably also a good deal of self-censorship. The result is that sites apparently offering a glimpse of public opinion are skewed representations. Unlike a newspaper that can be read in certain ways based on the assumption that it is censored and its content controlled, websites give the impression of consisting simply of the spontaneous postings of the public. This doesn't mean that the web doesn't bring a gain in sharing information. Indeed, journalists on official Chinese newspapers have been known to anonymously post to the web stories censors would stop them from publishing—and these get some airing even if they are quickly censored. And, to complicate the matter further, the government sometimes tolerates web content that it would censor from print publication. The reason seems to be that it assumes that the Internet reaches a relatively elite part of the population within which it is willing to see issues aired. This may offer a kind of safety valve; it may be monitored by the government to see what its more publicly informed and vocal citizens think; it has the advantage that it doesn't seem as official as what appears in government-owned publications. Perhaps most basically, the government may simply recognize that it cannot effectively restrict all information from reaching this elite—but still wish to restrict what reaches the broader population. In the information age, it may be all but impossible to stop information flows—including across borders—but it may be more possible to keep them relatively elite. In other settings, censorship may ultimately be less of a factor in determining the accessibility of information on the web than invisibility (because of the limits of search engines, the extent to which commercial sites push public ones to the margin, and the costs of advertising or maintaining links).

One argument has been that web-based "communities" are effective organizational counterparts to corporations, governments, and other more formally organized structures of power and action. There is some truth to this—organization is always crucial to making otherwise dispersed popular voices heard. Electronic communication is extremely effective at facilitating links among strangers—for example facilitating both widely distributed protests and participation from widely distributed groups in demonstrations or other shared events. It is a useful vehicle for sharing information which can either motivate or inform action. At the same time, we need to note the ambiguity of the language of community in this regard. Writers seem to refer without clear distinction to (a) dense networks of personal ties, (b) broad but loosely linked categories of people linked by shared information, and (c) IT users who derive their social connections from some other basis (say, trade union membership) but then communicate partly online. If we mean the first of these, the most sociologically specific use of 'community', then communities are effective only on relatively small scales; density of connections—everyone knowing (and potentially trusting) everyone else necessarily declines with scale. Precisely because social life began to be organized on ever-larger scales in the modern era, community could organize proportionately less of it.

States, markets, and other kinds of connections among people came to organize more— even when people still valued community a great deal. The term community is misleading, moreover, because it implies that collectivities formed in web-based discourse are somehow equivalent to local communities grounded in face-to-face relationships. Here too the reality is that the Internet is most effective as a complement to face-to-face communication not a replacement for it. It empowers local activists who would otherwise find it harder to reach others with similar concerns in remote locations. It enables both lateral sharing of information and better access to information controlled by centralized providers (including government agencies). Here it is worth noting that how effective the technology is in this regard depends a great deal on institutional arrangements over which there can be political struggle. Are government agencies required to make all their data effectively accessible, for example? What laws and regulations govern public provision of data by other actors?

More generally, "community" is a misleading term for thinking about the Internet's role in social solidarity. It may be true, as William Mitchell puts it, that his keyboard is his café. Internet communications provide many of the functions that coffee houses (and newspapers) played in Habermas's model of the 18th century public sphere. At the same time, neither is necessarily best understood through the notion of community. Part of what Habermas described as important to coffee house society was precisely that it provided a setting for interactions among strangers, for discussions among those who were not knit to each other by communal ties. The same goes for various sorts of multi-user domains and is part of the basis for the creative identity-work that Sherry Turkle has described. MUDs may evoke a sense of community for participants, and may be given metaphorical topographies complete with "neighborhoods". But it is also misleading to apply the warm, fuzzy language of the local without regard to scale. There is a crucial difference among MUDs that provide occasions for strangers to fantasize, LISTSERVs and websites that enable political organization among strangers, and web-resources that provide supplemental communication channels for people also joined in directly interpersonal relationships. The three (and other variants) can shade into each other, but are distinct not just in conception but in implications. A particularly important part of the distinction comes with the difference between categorical identifications (categories of people with similar interests or identities) and network ties (more or less dense webs of actual relationships, especially multiplex ones). The latter join more or less "whole" people; the former tend to join people one issue or identity at a time. The former, however, transcend locality much more easily.

The point is not that one mode of interconnection is better or worse than the other, but that they have different implications. The Internet can make local community less isolated or it can lead people to substitute on-line ties to relative strangers for interaction with neighbors. It can promote both enclaves and connectivity, nations and cosmopolitanism. Heavy reliance on the term 'community' to describe computer-mediated groupings borrows from the warm and fuzzy connotations that the idea of community has in everyday life and especially in nostalgia. But it also obscures one of the most important potential roles for electronic communication which is enhancing public discourse—a form of discourse that joins strangers and enables large collectivities to make informed choices about their institutions and their future.

The Idea of the Public Sphere

To say that a public exists is mainly to say that there is more or less open, self-organized communication among strangers. There are settings in which such public communication is minimal, in which strangers feel sharply constrained in what they can or should say to each other, or in which external regimentation of who may speak to whom or what may be the topics of communication prohibits self-organization of publics. So this minimal sense of publicness says a good deal. It does not, however, give an adequate sense of the stakes of the idea of the public sphere. These focus not simply on the general existence of public communication, but on its capacity to guide social life.

The most famous study of the public sphere, Jurgen Habermas's *Structural Transformation of the Public Sphere*, focuses on the emergence in the 18th century of a widespread ideal—and partially successful actual practice—of open debate concerning questions about the public good and policies for pursuing it. This debate was conducted by individuals with autonomous bases in civil society—i.e., with their own sources of income and identity—but the debate achieved rationality to the extent that the wealth and identities of the individuals did not intrude on argument itself. The public sphere was thus crucial to the capacity of civil society to influence the state; it was political, but not part of the state. Habermas describes the development of this political public sphere out of an earlier literary one, in which the formation of opinion about works of literature was elevated by rational-critical discourse. He also points to the importance of institutional supports for public communication, notably the newspaper and the coffee house.

In Habermas's analysis, however, the ideal of the public sphere was shown to mask a contradiction. It aspired equally to openness—in manner of communication and in entrance to the discourse itself—and to the critical use of reason to form opinion. Its successive structural transformations, however, reflected the extent to which expanding the scale of the public sphere led to a degeneration in the processes of opinion formation. Instead of individuals debating, public communication became increasingly a matter of organized interest groups—corporations, trade unions, political parties—using the techniques of advertising and mass communications. Opinion was formed less rationally on the basis of manipulation. The advent of opinion polls did nothing to arrest this, since they simply asked the opinion of discrete individuals without providing an occasion for different individuals to inform each other through discussion. Such opinions were less likely, therefore, to be based on knowledge. Indeed, opinion polls simply gave better information on people to those who would seek to influence or control them—treating them as "objects"—rather than enabling people to become better subjects of public discourse.

Behind both Habermas's optimistic account of the 18th century and his pessimistic account of subsequent transformations lie the same issues. Can public communication be a means of forming opinions based on reason and improved by critical discourse? Can these opinions then inform the creation or operation of social institutions and more generally the constitution of social life. In other words, can ordinary people use their faculties for reason and communication to choose the nature of their lives together—even on a relatively large scale in which most are strangers to each other—rather than have this imposed on them by inherited tradition, political power, or economic wealth. For

Habermas, the main issue was whether citizens could guide a state, which helps to explain why "the public sphere" appears as singular—it is the sum of the ways in which the open, rational-critical discourse of citizens at large can inform the (singular) state. The same broader questions can be asked of international public discourse, however, and with regard to influencing a multiplicity of states, of multilateral organizations, of quasi- or non-governmental bodies, and indeed, of business corporations.

Communication among strangers in the public sphere thus appears as one of the various mechanisms by which social integration is achieved and social institutions shaped and guided. Shared participation in public life, itself, can be a form of social solidarity, joining people in relations to each other and to the public projects they debate—even when they disagree about specifics. Among the various forms of social integration, public communication is distinctive because more than any of the others it offers the possibility of ordering collective life on the basis of reasoned choice. People are joined together by functional interdependence, as in markets, by subjection to political power, and by commonalties of culture. The public sphere has the potential to shape each of these other forms of connection, as well as offering a kind of connection of its own. It complements, the realms of directly interpersonal relationships—family, friendship, neighborhood. These are open in varying degree to choice and to construction through interpersonal communication. But they can join only modest numbers of people; they cannot organize relations among strangers. Moreover, when they steer affairs of state this is usually the illegitimate influence of one part of a polity over the rest. The public sphere distinctively offers the chance for the whole citizenry—or at least a large part of it—to participate in guiding the state, thus conferring legitimacy on it. This of course implies that legitimacy is based on the will of the people. While this is a predominant conception in the modern world, it is not the only possibility. Other approaches to legitimacy base it on divine will, on inheritance, or on the formal excellence of institutional arrangements. When legitimacy is based on the will of the people, though, a key question is how that will is formed. Is it a matter of inheritance or of a tradition that should be above question—as some proponents of ethnic nationalism would argue? Is it a matter of manipulation by those who can draw power from control over the media, or over jobs, or over the members of self-interested organizations? Or is this will of the people formed through discussion and debate among the people?

This is not the place to pursue questions of political theory like how to balance unity and diversity, the sense of common commitment implied by the phrase "will of the people" and the need for tolerance and mutual engagement across lines of difference required by liberal democracy. But it is crucial to see that these questions are made pressing by any appeal to the public sphere as a basis for organizing life together. Take simply the relationship to nation as an example. Nationalist ideology commonly presents individuals as joined directly and identically to a common whole; differentiation appears as a threat to national unity. Such an ideology can have popular appeal and can be supported by masses of public communication—messages posted to both American and Chinese websites during the spyplane crisis of 2000 demonstrated this. But this sort of public demonstration of unity and identity is different from public discourse in which rational-critical debate improves the quality of opinions, educates the participants and forms a collective understanding of issues that advances beyond pre-existing definitions of interests or identities. Among other things, the mass version of popular nationalism is eminently open to manipulation by demagogues, the government, and other forces. Recall, indeed, that democracy can mean mass rule not protection for rights and that publicity and expansion of the media can be tactics in mobilizing such masses.

More or less open, self-organized public communication among strangers is evident in both mass nationalism and more rational-critical public discourse. The minimal definition of the public sphere does not adequately make the distinction (though to the extent that mass nationalism is manipulated by the state or others it is not 'self-organized'). It is recognizing the deeper stakes of the question of the public sphere that makes the tension between these two "ideal types" of public communication evident and important. And yet, it should not be imagined that the realities to which they refer are completely distinct. There is no communication that is not also participation in the production or reproduction of culture. Even in the most rational-critical of public debates, thus, organized by the fairest rules of procedure and pursued in the least self-interested spirit, are thus conditioned by the cultural idioms in which they are conducted and contribute to the formation of common culture (as well as the recognition of cultural differences) among participants. Moreover, common culture is important to the commitment of participants to each other and to the process of public discourse. To recognize the difference between simple affirmation of commonalty and differentiated public debate should not lead us to imagine that the ideal of reasoned discourse is sufficient in itself to account for what goes on in any public sphere or how and why it survives and exerts its influence. Conversely, we should see also that popular nationalism (and other cultural or political traditions) is not immune from rational-critical discourse. It can be shaped and reshaped both by the culture-forming and the rational critical dimensions of public discourse. Even when ideologues assert that there is one true and correct way to be American or Chinese or anything else, they tend always to offer distinct visions of that one and true way. This opens the possibility of debate even in the midst of affirmations of unity. And different nationalisms may incorporate greater recognition of internal diversity, of the importance of reason, or of debate as part of the self-understanding of the nation.

The existence of public communication, vital as it is, does not answer the questions we need to ask about public spheres. We need to know not only how active the communication is and how inclusive and open participation is, but what the qualities of the communication are. We need to attend to the processes by which culture is produced and reproduced in public, not treat it as mere inheritance or private product of individuals or small groups. We need to ask how responsive public opinion is to reasoned argument, how well any public sphere benefits from the potential for self-correction and collective education implicit in the possibilities of rational-critical discourse. And we need to inquire how committed participants are to the processes of public discourse and through it to each other. Finally, and not least of all, we need to ask how effectively the public opinion formed can in fact influence social institutions and wielders of economic, political, or indeed cultural power.

These questions should all be basic to inquiry into the implications of new information technology for public life and democracy. Various technologies have the potential to constrain or facilitate openness, reason, cultural creativity, self-organization, solidarity. This is as true internationally as domestically. Moreover, questions of unequal access, cultural diversity, and perhaps most basically, the multiplicity of agents of power and potential objects of public influence loom even larger in the global arena.

The implications of information technology for the global public sphere are still being determined. Whether it will be put to use in ways that open and encourage public communication as much as in ways that facilitate commerce and control is an open question. If not, then web-based resistance to power— viruses, hacking, site-flooding, and other information technology and web-based strategies for attacking corporations, states, other users—may become more prominent. There is no perfect 'firewall' between systems of public communication and infrastructural systems. Accordingly, the stakes are high should cyberterrorism become more prominent. At the same time, opportunities for democratic choice about how sociotechnical systems are organized are limited. And it is often the sense of being excluded from choices about the future than fans the flames of discontent.

Prospects for democratization of the global order depend crucially on the development of a global public sphere, as well as on attention to global inequities within the public discourse of individual societies. There are opportunities for activism. This need not always involve a leap to the truly global, but may involve the building of transnational communications networks on a regional scale—as has happened to some extent around Latin America and South Asia. It will be important for those who would open up the public sphere to figure out how to work within organizations (including corporations and states) not just against them or in seemingly separate and autonomous "communities". It may also be possible to *choose* at least some aspects of the transformation of colleges and universities instead of just letting it happen, and thus to choose to have stronger bases for producing and sharing knowledge to inform the public. In this regard, one of the most important actions may be to do real research to help replace the contest of anecdotes and speculations with a reasoned debate in the public sphere.

References

Arendt, H. (1958) *The Human Condition*. Chicago: University of Chicago Press.

Calhoun, C. (ed.) (1992a) *Habermas and the Public Sphere.* Cambridge, MA: MIT Press.

Calhoun, C. (1992) 'The Infrastructure of Modernity: Indirect Relationships, Information Technology, and Social Integration', in H. Haferkamp and N. J. Smelser (eds), *Social Change and Modernity*. Berkeley: University of California Press.

Castells, M. (1996) *The Rise of the Network Society* (vol.1 of *The Information Age*). Cambridge, MA: Blackwell.

Mitchell, W. (1998) *City of Bits*. Cambridge, MA: MIT Press.

Wellman, B. and Haythornthwaite, C. (eds) (2003) *The Internet in Everyday Life*. Cambridge, MA: Blackwell.

White, H. (1992) *Identity and Control.* Princeton: Princeton University Press.

Yang, G. (2003) 'The Internet and the Rise of a Transnational Chinese Cultural Sphere', *Media, Culture & Society*, 25(4): 469—90.

3

Nature of Electronic Governance

E-Governance is the application of Information and Communication Technology (ICT) for delivering government services, exchange of information communication transactions, integration various stand-one systems and services between Government-to-Citizens (G2C), Government-to-Business(G2B),Government-to-Government(G2G) as well as back office processes and interactions within the entire government frame work. Through the e-Governance, the government services will be made available to the citizens in a convenient, efficient and transparent manner. The three main target groups that can be distinguished in governance concepts are Government, citizens and businesses/interest groups. In eGovernance there are no distinct boundaries.

Generally four basic models are available:

1. Government to Customer,
2. Government to Employees,
3. Government to Government and
4. Government to Business.

The terms "e-governance" and "e-government" are treated to be the same, however, there is some difference between the two. "E-government" is the use of the ICTs in public administrations- combined with organisational change and new skills- to improve public services and democratic processes and to strengthen support to public". The problem in this definition to be congruent with the definition of e-governance is that there is no provision for governance of ICTs. As a matter of fact, the governance of ICTs requires most probably a substantial increase in regulation and policy- making capabilities,with all the expertise and opinion-shaping processes among the various social stakeholders of these concerns. So, the perspective of the e-governance is "the use of the technologies that both help governing and have to be governed".

E-Governance is the future, many countries are looking forward to for a corruption free government. E-government is one-way communication protocol whereas E-governance is two-way communication protocol. The essence of E-governance is to reach the beneficiary and ensure that the services intended to reach the desired individual has been met with. There should be an auto-response system to support the essence of E-governance, whereby the Government realizes the efficacy of its governance. E-governance is by the governed, for the governed and of the governed.

Establishing the identity of the end beneficiary is a true challenge in all citizen-centric services. Statistical information pubslished by governments and world bodies do not always reveal the facts. Best form of E-governance cuts down on unwanted interference of too many layers while delivering governmental services. It depends on good infrastructural setup with the support of local processes and parameters for governments to reach their citizens or end beneficiaries. Budget for planning, development and growth can be derived from well laid out E-governance systems.

E-Government should have an intranet for ensuring smoother flows of data, communications and access to information by different Ministries and Departments. It should be made mandatory for all Ministries and other Departments to put as much information as possible on to the intranet but in a manner in which searches can be made easily. Transactions between various departments of the government of India and other Government organisations should be networked so that a substantial part of transfer of files and paper can be replaced by an intranet within the Government.. All departments & agencies to ensure that they operate web sites which provide upto date information, forms, leaflets etc. E-mail to be incorporated into the normal range of contact methods and departments & agencies will implement arrangement for rapid response to e-mail queries.

Service-wide guidelines and practices to be adopted regarding content format and presentation for web sites and inter-departmental group to be established to deal with these issues.MIT should bring out periodic bulletins on the Pilot project implementation in order to facilitate dissemination exchange of information. There should be a single web based front-end for all government services to the public. To use Internet and WWW based applications, Domain Name standardisation of various states must be done as it is available for various countries.

POWER OF E-GOVERNANCE

The term Digital Governance refers to governance processes in which Information and Communications Technology (ICT) play a significant role. The role played by ICT could be wide-ranging: *in delivery and standards of governance services, to how people access such services, and the participation of people in the governance sphere*. Digital Governance uses ICT to induce changes in the delivery and standards of governance services and more importantly, in the way citizens interact and participate in the governance sphere.

Access to information is power. The traditional power-bearers in the society have always realized the role of information to gain control and set up governance mechanisms in their constituencies.

In case of *good governance*, information is acquired and used strategically for *public good* purposes. And in case of *bad governance*, the same information is used for *private gains* and for suppression of the citizens. In either forms of governance, the selective use of information creates hierarchal structures on which power gets unequally distributed. The skew in the distribution of power at each hierarchal level is proportional to the critical information residing at each of these levels.

Some facts on the role of information in governance are:

1) Access to information forms the basis of decision-making (which could be for *public good* or *private gains*)
2) Well-informed decision-making is dependent on the quality and timeliness of information available
3) Limiting access to information to an exclusive group opens up avenues for manipulation of information for exploitative purposes, such as private gains

Good Governance

Good Governance rests on the pillars of knowledge and recognition of this set of knowledge by the decision makers and people alike. Digitization of this entire set of knowledge within a network which is open to all individuals (an *inclusive* network by design) opens up possibilities for all to access and use this knowledge— *paving the way for Digital Governance or E-governance as it is more popularly known.*

Introduction of Digital Governance ensures that citizens can participate in, and influence decision-making processes which affect them closely. Citizens no longer remain passive recipients of governance services provided to them, but can pro-actively decide the types and standards of governance services they want and the governance structures which can best deliver them.

ICT can influence governance processes possibly in 3 ways:

1) *Technical role*: Automation of repetitive governance tasks and thereby improving efficiency of governance processes. For instance, automated filing of tax forms, e-voting, periodic information reporting etc.
2) *Supportive role*: Use of ICT to complement existing efforts and processes to improve governance. For instance, use of Internet to catalyse existing efforts towards transparency in government information and functioning, or embedding use of emails in connecting decision-makers with their constituencies.
3) *Innovative role*: Use of ICT to initiate new governance services or new mechanisms for improved service delivery which would be impossible through non-ICT modes. For instance, online checking of status of an application (from remote and beyond office hours); providing instant access to the same information to all individuals through emails and website; ability to instantly access, compare or triangulate information from outside of the constituency or government sources etc.

Digital Governance

Digital Governance in Developing Countries Does Not Imply

Linking every citizen to a digital node or giving them access to Internet and computers.

Digital Governance in Developing Countries Implies

Ensuring every community or a village has easy access to information available on the digital network and no one is excluded from accessing information on this network. Access to information could be through:

1) Private/individual ICT nodes such as individual access to Internet
2) Public ICT nodes such as community Internet centre, post offices, public phone booths, government information centres
3) Convergence of modes such as extension volunteers, community radios and local newspapers which have access to an ICT node and can then relay critical information to targeted users in a timely manner. (implying convergence of ICT with conventional modes)

The use of Digital Governance transforms governance from "representative" to a more "individual based" form, and from "passive" to being "pro-active". It does not require an individual to be based in the local constituency "in-situ" to influence or benefit from governance delivery services. Further as use of digital governance leads to closer contact of individuals with decision-makers/officials in the government, the impact is immediate. On the whole, it puts greater access and control over governance mechanism in the hands of individuals, and in process leads to more transparent, accountable and efficient governance.

E-Government: Nature and Functions

E-Government is a diffused neologism used to refer to the use of information and communication technology to provide and improve government services, transactions and interactions with citizens, businesses, and other arms of government.

Delivery Models and Activities of E-Government

The primary delivery models of e-Government can be divided into:

1) Government-to-Citizen or Government-to-Consumer (G2C)
2) Government-to-Business (G2B)
3) Government-to-Government (G2G)
4) Government-to-Employees (G2E)

Within each of these interaction domains, four kinds of activities take place:

1) pushing information over the Internet, e.g: regulatory services, general holidays, public hearing schedules, issue briefs, notifications, etc.
2) two-way communications between the agency and the citizen, a business, or another government agency. In this model, users can engage in dialogue with agencies and post problems, comments, or requests to the agency.

3) Conducting transactions, e.g: lodging tax returns, applying for services and grants.

4) Governance,e.g: online polling, voting, and campaigning.

While e-government is often thought of as "online government" or "Internet-based government," many non-Internet "electronic government" technologies can be used in this context. Some non-Internet forms include telephone, fax, PDA, SMS text messaging, MMS, wireless networks and services, Bluetooth, CCTV, tracking systems, RFID, biometric identification, road traffic management and regulatory enforcement, identity cards, smart cards and other Near Field Communication applications; polling station technology (where non-online e-voting is being considered), TV and radio-based delivery of government services, email, online community facilities, newsgroups and electronic mailing lists, online chat, and instant messaging technologies.

Risks

There are many considerations and potential implications of implementing and designing e-government, including disintermediation of the government and its citizens, impacts on economic, social, and political factors, vulnerability to cyber attacks, and disturbances to the *status quo* in these areas.

Hyper-surveillance

Increased contact between government and its citizens goes both ways. Once e-government begins to develop and become more sophisticated, citizens will be forced to interact electronically with the government on a larger scale. This could potentially lead to a lack of privacy for civilians as their government obtains more and more information on them. In a worse case scenario, with so much information being passed electronically between government and civilians, a totalitarian-like system could develop. When the government has easy access to countless information on its citizens, personal privacy is lost.

Cost

Although "a prodigious amount of money has been spent" on the development and implementation of e-government, some say it has yielded only a mediocre product. The outcomes and effects of trial Internet-based governments are often difficult to gauge or unsatisfactory.

Inaccessibility

An e-government site that provides web access and support often does not offer the "potential to reach many users including those who live in remote areas, are homebound, have low literacy levels, exist on poverty line incomes, suffer from chronic illness, are single parents or older adults."

False Sense of Transparency and Accountability

Opponents of e-government argue that online governmental transparency is dubious because it is maintained by the governments themselves. Information can be added or removed from the public eye (i.e. the Internet) with or without public notice. For example, after the World Trade Center in New York City was attacked on September 11, 2001, United States federal officials removed a large amount of government information from its websites in the name of national security. This act went relatively unnoticed by United States citizens. To this day, very few organizations monitor and provide accountability for these modifications. Those that do so, like the United States' OMBWatch and Government Accountability Project, are often nonprofit volunteers. Even the governments themselves do not always keep track of the information they insert and delete.

Benefits

It is convenient and cost-effective for businesses, and the public benefits by getting easy access to the most current information available without having to spend time, energy and money to get it.

E-government helps simplify processes and makes access to government information more easily accessible for public sector agencies and citizens. For example, the Indiana Bureau of Motor Vehicles simplified the process of certifying driver records to be admitted in county court proceedings. Indiana became the first state to allow government records to be digitally signed, legally certified and delivered electronically by using Electronic Postmark technology. In addition to its simplicity, e-democracy services can reduce costs. Alabama Department of Conservation & Natural Resources, Wal-Mart and NICdeveloped an online hunting and fishing license service utilizing an existing computer to automate the licensing process. More than 140,000 licenses were purchased at Wal-Mart stores during the first hunting season and the agency estimates it will save $200,000 annually from service.

The anticipated benefits of e-government include efficiency, improved services, better accessibility of public services, and more transparency and accountability.

One goal of e-government will be greater citizen participation. Through the internet, people from all over the country can interact with politicians and make their voices heard. Blogging and interactive surveys will allow politicians to see the views of the people they represent on any given issue. Chat rooms can place citizens in real-time contact with elected officials and their offices, allowing voters to have a direct impact and influence in their government. These technologies can create a more transparent government, allowing voters to immediately see how and why their representation in the capital is voting the way they are. This helps voters better decide who to vote for in the future. A government could theoretically move more towards a true democracy with the proper application of e-government. Government transparency will give insight to the public on how decisions are made and hold elected officials accountable for their actions. The public could become a direct and prominent influence in government legislature to some degree.

Environmental Bonuses

Proponents of e-government argue that online government services would lessen the need for hard copy forms. Due to recent pressures from environmentalist groups, the media, and the public, some governments and organizations have turned to the Internet to reduce this paper use. The United States government utilizes the website http://www.forms.gov to provide "internal government forms for federal employees" and thus "produce significant savings in paper.

Speed, Efficiency, and Convenience

E-government allows citizens to interact with computers to achieve objectives at any time and any location, and eliminates the necessity for physical travel to government agents sitting behind desks and windows. Improved accounting and record keeping can be noted through computerization, and information and forms can be easily accessed, equaling quicker processing time. On the administrative side, access to help find or retrieve files and linked information can now be stored in databases versus hardcopies stored in various locations. Individuals with disabilities or conditions no longer have to be mobile to be active in government and can be in the comfort of their own homes.

Public Approval

Recent trials of e-government have been met with acceptance and eagerness from the public. Citizens participate in online discussions of political issues with increasing frequency, and young people, who traditionally display minimal interest in government affairs, are drawn to e-voting procedures.

Although internet-based governmental programs have been criticized for lack of reliable privacy policies, studies have shown that people value prosecution of offenders over personal confidentiality. Ninety percent of United States adults approve of Internet tracking systems of criminals, and fifty-seven percent are willing to forgo some of their personal internet privacy if it leads to the prosecution of criminals or terrorists.

There are also some technology-specific sub-categories of e-government, such as m-government (mobile government), u-government (ubiquitous government), and g-government (GIS/GPS applications for e-government.

E-government portals and platforms The primary delivery models of e-Government are classified depending on who benefits. In the development of public sector or private sector portals and platforms, a system is created that benefits all constituents. Citizens needing to renew their vehicle registration have a convenient way to accomplish it while already engaged in meeting the regulatory inspection requirement. On behalf of a government partner, business provides what has traditionally, and solely, managed by government and can use this service to generate profit or attract new customers. Government agencies are relieved of the cost and complexity of having to process the transactions.

To develop these public sector portals or platforms, governments have the choice to internally develop and manage, outsource, or sign a self-funding contract. The self-funding model creates portals that pay for themselves through convenience fees for certain e-government transactions, known as self-funding portals. Early players in this space include govONE Solutions, First Data Government Solutions and NIC, a company built on the self-funded model.

Social networking is an emerging area for e-democracy. The social networking entry point is within the citizens' environment and the engagement is on the citizens' terms. Proponents of e-government perceive government use of social networks as a medium to help government act more like the public it serves. Examples of state usage can be found at The Official

Commonwealth of Virginia Homepage, where citizens can find Google tools and open social forums.

Government and its agents also have the opportunity to follow citizens to monitor satisfaction with services they receive. Through ListServs, RSS feeds, mobile messaging, micro-blogging services and blogs, government and its agencies can share information to citizens who share common interests and concerns. Government is also beginning to Twitter. In the state of Rhode Island, Treasurer Frank T. Caprio is offering daily tweets of the state's cash flow. Interested people can sign up at here. For a full list of state agencies with Twitter feeds, visit Real Life. Live document.

There are several international rankings of e-government maturity. The Eurostat rankings, Economist, Brown University, and the UN e-Government Readiness Index are among the most frequently cited. The United Nations conduct an annual e-Government survey which includes a section titled *e-Government Readiness*. It is a comparative ranking of the countries of the world according to two primary indicators: i) the state of e-government readiness; and ii) the extent of e-participation. Constructing a model for the measurement of digitized services, the Survey assesses the 191 member states of the UN according to a quantitative composite index of e-government readiness based on website assessment; telecommunication infrastructure and human resource endowment.

Public information in Canada is the subject of the Access to Information Act. VisibleGovernment.ca is a Canadian non-profit that promotes online tools for government transparency. There have been several ChangeCamps in Toronto, Ottawa and Vancouver, with organizers coming together in Edmonton, Halifax, Montreal, drawing many politicians. The city of Toronto mayor David Miller announced plans for an open city data portal at toronto.ca/open.

The election of Barack Obama as President of the United States has become associated with the effective use of Web 2.0 technologies during his campaign, and in the implementation of his new government in 2009.

On January 21, 2009, newly elected President Obama signed one of his first memorandums—the Memorandum for the Heads of Executive Departments and Agencies on Transparency and Open Government. In the memo, President Obama called for an unprecedented level of openness in Government, asking agencies to "ensure the public trust and establish a system of transparency, public participation, and collaboration." The memo further "directs the Chief Technology Officer, in coordination with the Director of the Office of Management and Budget (OMB) and the Administrator of General Services (GSA), to coordinate the development by appropriate executive departments and agencies, within 120 days, of recommendations for the Open Government Directive, to be issued by the Director of OMB, that instructs executive departments to take specific actions implementing the principles set forth in the memorandum." Common interpretation of this memorandum is that Obama, whose presidential campaign was heavily influenced by web 2.0 technology, is calling for the utilization of web 2.0 technology across all federal departments and agencies.

e-Governance: The Indian Experience

Global shifts towards increased deployment of IT by governments emerged in the nineties, with the advent of the World Wide Web. What this powerful means to publish multimedia, support hyperlinked information and interactive information meant was a clearer avenue for G to C interactions and the promise of the attainment of the goals of good governance. Governments weighed down by the rising expectations and demands of a highly aware citizenry suddenly began to believe that there can be a new definition of public governance characterized by enhanced efficiency, transparency, accountability and a citizen-orientation in the adoption of IT enabled governance.

E-governance originated in India during the seventies with a focus on inhouse government applications in the areas of defence, economic monitoring, planning and the deployment of ICT to manage data intensive functions related to elections, census, tax administration etc. The efforts of the National Informatics Center (NIC) to connect all the district headquarters during the eighties was a watershed. From the early nineties, e-governance has seen the use of IT for wider sectoral applications with policy emphasis on reaching out to rural areas and taking in greater inputs from NGOs and private sector as well. There has been an increasing involvement of international donor agencies such as DfID, G-8, UNDP, WB under the framework of e-governance for development.

While the emphasis has been primarily on automation and computerization, state endeavours to use IT include forays into connectivity, networking, setting up systems for processing information and delivering services. At a micro level, this has ranged from IT automation in individual departments, electronic file handling, access to entitlements, public grievance systems, service delivery for high volume routine transactions such as payment of bills, tax dues to meeting poverty alleviation goals through the promotion of entrepreneurial models and provision of market information. The thrust has varied across initiatives, with some focusing on enabling the citizen-state interface for various government services, and others focusing on bettering livelihoods

India is a land of diversity. This diversity spans across culture, tradition, language, geography and the economic condition of the people. It is a nation that has a significant number of people who are below the minimal socio-economic benchmarks. This includes rural and urban poor, women in rural areas, street children, people belonging to historically disadvantaged castes and people living in less developed areas. The vulnerability of these sections of society has increased with globalisation and this section is prone to become even more marginalized—economically and socially.

Successive governments have committed themselves to addressing these divides, but effective implementation of various economic development programmes aimed at individuals belonging to these sections of society has proved an elusive goal. Government of India (GoI) recognizes that e-Governance, in the context of developing countries, provides an excellent opportunity for improving governance. Used imaginatively, it is a trigger for introducing various administrative reforms. These changes could not only go a long way in improving the quality of life of these sections of society, but could actually provide them more equitable access to economic opportunities than ever before. In this context, the Government of India views e-Governance as a strategic tool for transforming Governance and improving the quality of services provided by the government to its people.

India's experience in e-Governance / ICT initiatives has demonstrated significant success in improving accessibility, cutting down costs, reducing corruption, extending help and increased access to un-served groups. In this phase of *experimentation*, e-government initiatives have reached millions of people belonging to these sections of society. Improved access to information and services has provided economic and social development opportunities, facilitated participation and communication in policy and decision-making processes and empowerment of the weakest groups. This has led to fostering a sense of ownership and building of social capital, which in turn, constitute a basis for local revitalization.

The Government of India, in various forums, has indicated its commitment to provide efficient and transparent government to all strata of society. E-Governance is now mainly seen as a key element of the country's governance and administrative reform agenda. The Government of India aspires to provide:

— Governance that is easily understood by and accountable to the citizens, open to democratic involvement and scrutiny (an open and transparent government)

— Citizen-centric governance that will cover all of its services and respect everyone as individuals by providing personalised services.

— An effective government that delivers maximum value for taxpayers' money (quick and efficient services)

Hence the Government of India views e-Governance as a vehicle to initiate and sustain reforms by focusing on three broad areas:

Initiatives across the country. Indicated below are some of the key initiatives taken in the country across some of the important citizen/business related departments:

1. Customs and Excise (Government of India)
   — 98% of export and 90-95% of import documentation computerized
   — Electronic filing through ICEGATE at 3 locations (Mumbai, Delhi, Chennai)
   — 80% of Service Tax returns electronically processed
2. Indian Railways (Government of India)
   — Anywhere to Anywhere reservation from Anywhere
   — Electronic Booking of tickets on select sectors
   — Online Information on Railway reservation on Internet
3. Postal Department (Government of India)
   — Direct e-credit of Monthly Income Scheme returns into the investors accounts
   — Dematerialization of Savings Certificate (NSC) and Vikas Patras (KVP), offering full portability
4. Passport / Visa (Government of India)
   — 100% passport information computerized

— All 33 Regional Passport Offices covered

— Machine readable passports at some locations

5. AP Online (State Government of Andhra Pradesh)

An Integrated Citizen Services Portal providing citizen centric services such as: Birth/Death Certificates, Property Registration, Driver's License, Govt. Applications & Forms, Payment of taxes / utility bills etc.

6. Bhoomi – Automation of Land Records (State Government of Karnataka)

It provides computerized Record of Rights Tenancy & Crops (RTC)—needed by farmer to obtain bank loans, settle land disputes etc. It has also ensured increased transparency and reliability, significant reduction in corruption, exploitation and oppression of farmers. This project has benefited 20 million rural land records covering 6.7 million farmers.

7. CARD – Registration Project (State Government of Andhra Pradesh)

Computerisation Administration of Registration Department (CARD) impacting 10 million citizens over a period of 3 years. It has completed registration of 2.8 million titles with title searches made in 1.4 million cases. The system ensures transparency in valuation of property and efficient document management system. The estimated saving of 70 million man-hours of citizen time valued at US$ 35 mil (investment in CARD—US$ 6million). Similar initiatives in other states like SARITA (State Government of Maharashtra) STAR (State Government of Tamil Nadu), etc. have further built upon this initiative.

India has recognized the benefits of e-governance and through it, ushered in a paradigm of citizen-centric service delivery. E-governance is reforming the way government manages and shares information with external and internal clients. Specifically, it harnesses information and communications technologies (such as Wide Area Networks, the Internet, and mobile computing) to transform relations with citizens, businesses and amongst various arms of government.

There are however, numerous challenges. Some of the key areas needing attention are:

— *Clarity in objective setting:* Project approval and funding of projects through multiple departmental budgets lead to wide variations in the approach to project objective setting, without a clear focus on outcomes or on building sustainable services. The service needs of citizens/ businesses and those of other departments are often either overlooked or accorded lower priority in relation to internal needs. Very often, objective setting is purely in ICT terms such as computers, networks and so on which are specified in great detail, while government business process outcomes are either not defined or are defined in vague terms that do not lend themselves to measurement post implementation.

— *Ensuring service delivery*: E-governance projects have primarily focused on internal process automation and generally are hardware and infrastructure driven (sometimes even vendor driven) with little focus on citizen service delivery or outcomes.

— *Leveraging Private Capital*: Experience of successful e-governance initiatives indicates that well structured service-oriented projects can attract private capital linked to explicit service-linked revenues from users or from government. The current system of project formulation, i.e. based on budgetary allocation / grant places little or no pressure on departments to develop project structures that can attract private capital, a goal that necessitates additional rigour and complexity at the project formulation and development stage.

— *Need for Government Process Reengineering*: Computerization of inefficient processes can lead to higher rather than lower levels of inefficiency and spiralling cost. Hence, it is essential to undertake process re-engineering as an integral part of e-governance project implementation in order to ensure increased efficiency and reduced costs.

— *Standardization:* Departmental approach and absence of a national framework for common standards has resulted in adoption of different technical standards and varied architectures. This has significant implications for designing effective integrated applications and also entails long-term costs and sub-optimal results.

— *Independent Impact Assessment*: In the current system, there is no requirement or institutional mechanism for an independent assessment of projects post-implementation to determine whether they have achieved the set objectives, except in purely financial terms. Further, very few projects have formal performance metrics defined at the start of the project to measure outcomes.

— *Localisation:* ICT solutions were mostly developed with an English Language interface. However, in India a vast majority (95%) of the citizens do not know English and use the local language. The fact is that India has 22 official languages; for success of e-Governance, this reality needs to be reflected in the implementation strategy.

— *Internal Capacity/Project Management Expertise*: Departments/ states have limited access to any institutional mechanisms for building capacities in the areas of e-governance project development and design, bid process management, professional project management, development of Contractual Frameworks, Service Level Agreements, etc. They also have no means of accessing external (outside Government) competencies that are difficult to procure within the governmental framework.

National E-governance Plan

Taking note of the potential of e-governance to improve the quality of life of the vast population of the country, the Government of India has formulated a national program – the National e-governance Plan (NeGP). This plan attempts to cover all the important areas relating to e-Governance – Policy, Infrastructure, Finances, Project Management, Government Process Reengineering, Capacity Building, Training, Assessment and Awareness etc. across the Central and State Governments.

NeGP is a comprehensive "programme" of the Government of India and is designed to leverage capabilities and opportunities presented by ICT to promote good governance across the country. One of the learning that is at the core of the NeGP is the emphasis on implementation of such projects with clear timelines and responsibility allocations – in a "Mission Mode". The Plan initially extends over a 4-year period at an estimated cost of over USD 3 billion. NeGP is aimed at introducing e-Governance systematically through 25 Mission Mode projects, which would touch the lives of more than 1 billion people.

The vision of NeGP is to make all Government services accessible to the common man in his locality through common service delivery outlets. The implementation strategy envisages clear definition of service goals and metrics for each project and structured stakeholder consultations with all stakeholders including citizens and civil society organizations before the service goals of each project are firmed up. Even at the stage of formulation of the NeGP, its vision and proposed strategy, consultations were held with various stakeholders including state governments, ministries/departments, IT industry representatives and civil society organizations.

The focus of the plan includes the following sectors/projects:

Agriculture

E-governance projects in the agricultural sector can provide benefit to farmers and the rural people and also enhance the lives of urban poor. There are numerous sub-projects pertaining to provision of timely expert advice to farmers, food security, marketability and commercial information relating to agricultural products, enhancing crop productivity, enhancing the reach of and ease of access to micro-credit, etc.

Municipalities

The coordinating agency is the Ministry of Urban Development. The main programs relevant to vulnerable and marginalized groups are registration of births and deaths, grievances and suggestions, health programs, etc.

Gram Panchayats

The important programmes being implemented by the Ministry of Rural Development for poverty reduction are employment generation, provision of basic services, infrastructure development etc.

The objective is to increase participation of rural population in the government and women empowerment.

Common Service Centres (CSCs) is one of the integrated projects envisioned in NeGP. The CSCs provide assisted community access points – a necessity in a country with relatively low levels of literacy and ICT penetration in rural areas. These centres are very effective in providing multiple services provided by different departments at a single location. For a common citizen, it is often confusing and time-consuming to have to visit different departments and identify the right official or office to avail of some service. This one stop shop is also helpful in increasing accessibility, enabling faster service delivery, curbing corruption and reducing difficulties faced by vulnerable and marginalized groups. Under this program, it is aimed to establish 100,000 CSCs predominantly in the rural areas to serve the needs of the traditionally underserved areas.

Core Infrastructure such as State Wide Area Networks, Data Centres, Gateways, etc. forms another critical element of the NeGP. Approximately

15% of the total program outlay is earmarked for common core and support infrastructure that is shared across projects, excluding the cost of infrastructure that is created specific to, and as a part of, individual projects.

The diagram presents the E-Governance framework that includes back-ends (databases of different government agencies, service providers, state governments etc.), middleware and front-end delivery channels (home PCs, mobile phones, kiosks, integrated citizen service centers etc) for citizens and businesses. The middleware comprises of communication and security infrastructure, gateways and integrated services facilitating integration of inter-departmental services.

GoI has further planned to carry out Impact Assessments of such projects to see whether the benefit is actually percolating to the marginalized sections of society. The Impact Assessment would cover the following:

— *Policy Making* – The agenda for e-Governance is typically set at the highest level of Government. Thus, the capacity at the highest political and bureaucratic levels is extremely critical for informed policy making. This also helps in sustaining such programmes. Further as NeGP entails significant Government Process Reengineering (GPR), adequate capacity at these levels is necessary to catalyse and drive such decisions.

— *Institutional Arrangements*—Implementing e-Governance policies that are consistent with a broad policy and adhere to common standards despite being implemented by numerous state and central government departments requires setting up of appropriate and empowered institutional arrangements to oversee, drive and manage implementation. These arrangements would vary from Central to State Governments although there would consistency of key roles i.e. formulating and ensuring implementation of e-Governance policies, addressing implementation bottlenecks and dependencies and finally monitoring progress and desired outcomes. These institutional mechanisms include both full-time bodies and committees/ groups that meet on an episodic or periodic basis.

— *Access to Professional Expertise*—For any institutional arrangements to be effective, there needs to be not only enough manpower; the competencies put together should be suited to the roles envisaged. This is an area, which probably requires the maximum attention. The focus of the capacity building initiatives has been to source the right people for the right job so as to make existing and proposed institutions effective. It is recognized that all the expertise required is not present within government and cannot be created exclusively by retraining existing personnel. Government departments and ministries were created to perform certain functions. Implementing e-governance projects was not one of them. Yet, no e-governance project can be meaningfully implemented without the active participation and leadership of the departmental personnel themselves. Even though the NeGP envisages accessing skills outside the government, a certain minimum level and combination of skills within the government are essential. Hence a carefully calibrated mix of inducting professionals at various levels from the market and retraining of existing personnel has been built into the NeGP.

— *Outcome Monitoring* – It has been our experience that to make an e-Governance initiative sustainable and cost-effective, there needs to be enough benefit accruing out of the project, which is monitored closely. Wherever possible, there also needs to be an explicit linkage between the benefit and the financing of the initiative.

Levels of Capacity Building

As NeGP in the states cuts across departments, it is expected that the State Government would provide leadership, direction and vision for e-Governance at the highest political and administrative levels, which would lead to a broad roadmap and prioritisation of projects. To enable the State administration to carry out the requisite groundwork, prepare project proposals, implement the projects and oversee O&M thereafter, adequate support through a dedicated, professional teams need to be in place with appropriate skill-sets and aptitudes at two levels:

— Programme level (i.e. at the State Level)

— Project level (i.e. at Department Level)

The first level deals with issues that cut across projects such as overall policies, strategies, technologies, common infrastructure and so on. These issues need to be dealt with by a department designated by the State Government typically the State IT department / e-Governance department. The second level deals with all project level (i.e. departments of State Government initiating e-Governance projects) issues from conceptualisation to implementation and O&M phase. The issues involved here are considerably different and are intimately connected with the domain in which the department functions.

Further Capacity Building also helps to maintain continuity of approach despite changes of key incumbent officials. Training programmes as a part of capacity building have been designed to cater to various needs. These are briefly described below:

E-Gov Champions Programme

This programme is intended for political and administrative heads of departments/ organizations who may not be concerned with the details of an e-governance project, but who nonetheless have to provide the leadership and an environment conducive to make it happen. Since such persons are typically extremely busy people, a short weekend program has been designed to familiarise them with the main issues that a CIO has to deal with and key perspectives regarding their role, particularly in change management.

Chief Information Officers

The accumulated wisdom from experience is that an e-governance project that is not lead and driven by a senior person from within the government domain in which it functions, is unlikely to succeed. Most such people, however, do not have the skills to lead a project through all its developmental stages. Hence a comprehensive training program extending over a duration of 3-6 months has been designed for identified officials. Typically, such persons have a scientific or technical background or an aptitude for project management. After undergoing this training program, an official is expected to have all the skills necessary to implement an e-governance project from conceptualization to complete rollout stage. This program has been designed in collaboration with the Indian Institute of Management, Ahmedabad. The number of people given this kind of training is small, say about one hundred or so, over the entire country. However, their impact on accelerating spread of e-governanceis enormous.

Chief Technology Officers

Since most projects require extensive technical skills which CIOs as defined above may not have, key personnel drawn from technical bodies like the National Informatics Centre (NIC) or state government departments/ PSUs are proposed to be trained to impart the skills that they need to have other than purely technical skills. Such CTOs with good domain knowledge could also function as CIOs. Another effort in this area is to provide opportunities for technical personnel to undergo advanced training programs in security, architecture, standards and so on to supplement their skill set appropriately.

Users of IT Systems in Government Departments

Such personnel are trained specifically on the package developed for that organization. Since this involves performance of tasks that they were already performing manually before the advent of the computerised system, a short two-week training program is found to be quite adequate. Typically, such training is part of the project and is built into the project in terms of cost, timing and content. The number of people to be trained depends on the size of the organization and the activities covered by the e-governance project. Typically, it runs into several hundred per project.

External Users of Government IT Systems

In many projects, there are organized bodies of external users of specific systems. For example, in the Ministry of Company Affairs project ('MCA21'), Company Secretaries and Chartered Accountants acting on behalf of companies incorporated in India have to file all returns, applications and other documents electronically as a mandatory requirement. An extensive program of training, awareness building and sensitization was built into the project as a part of the change management plan. Most e-Governance projects have external stakeholders for whom a capacity building plan is essential for success of the project in terms of utilization.

General IT Training

As computers become more pervasive, it is necessary to impart training to existing employees in use of computers for general purposes like word processing, spreadsheet, power point, etc. Typically, a five to ten day program is imparted by organizations like the National Informatics Centre (NIC) or private training institutions. The number of people to be so trained is very large.

Requirements

To initiate any e-Governance project, the capacities required are much more comprehensive and holistic than a conventional Software Development Life Cycle (SDLC) project. E-Governance initiatives entail policy development/ refinement, Financial Management, Program Management and warrant a higher emphasis on Change Management. Apart from this, creation of

Institutional structure which brings continuity to a project. However traditionally these have not been areas of focus. Hence while initiating any e-Governance projects there is an urgent need to develop capacities in all these areas at all the levels of Governance – Centre, State and Local Bodies.

To address these issues, we have factored in the federal structure of the country. Hence there is a need to create suitable institutions/capacities at different levels.

1. Central Government, for
   — Overall Policy Development, direction and Financial Management
   — Program Management and Support for e-Governance at national level
   — Management and implementation of projects under the Central Government
2. State Government, for
   — Providing direction at the state level to the e-Governance initiatives from the highest office in the State Government
   — Providing full time knowledge support to the policy makers and implementation teams
   — Individual project management
   — Creation/Strengthening of State Training Institutions
3. Local Bodies
   — For having in-house capabilities to manage e-Governance initiatives
   — Training Programme
4. Citizens
   — Awareness Programme
   — Right To Information
   — Civil Society Organizations

Taking cognizance of the above requirements, GoI has taken the following steps:

Capacity Building at the Centre

The Prime Minster's Office is directly involved in driving NeGP implementation. Further an Apex Committee under the chairmanship of the

Cabinet Secretary was constituted for overseeing and providing policy directions for the implementation of the National e Governance Plan (NeGP), and ensuring inter ministerial coordination. The Department of Information Technology (DIT) services the Apex Committee with the Secretary, DIT serving as the Convenor of the Committee. It was also felt that the resources required to manage and monitor a programme of this magnitude just did not exist anywhere in the country and needed to be developed. Therefore, a Programme Management Unit (PMU) has been created to ensure that individual e-Governance initiatives are given a "Programmatic" shape and content, adhere to a common set of objectives, approach and philosophy.

Capacity Building in Public Private Partnerships

Considering the important role that Public Private Partnerships are expected to play in the unfolding e-governance agenda, the Government of India, jointly with the IT industry body NASSCOM, set up the National Institute of Smart Government (NISG) which is itself a Public Private Partnership with 51% industry/ Financial Institution stake. The stated mission of this entity is to lead private sector resources (financial, technical and manpower) in an orderly and structured manner into the national e-governance effort. The NISG undertakes project development, training for policy makers and CIOs and provides handholding support to departments during the initial stages of project conceptualization and implementation by combining the best of resources from the government and the private sector. The NISG has also helped set up the PMU for the NeGP.

Capacity Building at the State Government

There is a huge variation in the capacities of various State Governments in the country to handle e-Governance. The size of the IT departments across the 28 states varies from a nucleic structure to a large full-fledged department.

As part of the capacity building initiative at the State Level, NeGP proposes the formation of dedicated institutionalised full time task forces at the state level and respective departmental levels called State e-Governance Mission Team (SeMT) and Project e-Governance Mission Team (PeMT) respectively. The approach to Capacity Building starts with formulating a detailed e-Governance roadmap for the State Government and deriving from it the expertise required to achieve the goals laid out.

Institutional Framework at State Level

An Apex Council headed by the Chief Minister (elected Chief Executive of the State Government) steers the entire e-Governance Program of the State at the highest level. This steering committee is indicative of the political will of the state administrative machinery to champion the cause of e-Governance and helps in synergizing efforts across the state to enable ICT emerge as a core enabler of good Governance in the state.

A State Apex Committee headed by the Chief Secretary addresses all operational issues. It consists of Secretaries of various departments. The Committee oversees and provides policy direction to NeGP & ensures inter-departmental co-ordination at the state level. It is expected that all Secretaries of departments implementing Mission Mode projects under NeGP are members of this committee.

SeMT- is a dedicated body at the State level consisting of full-time experts to provide an overall direction, standardisation and consistency through program management of e-Governance initiatives under way in the State Government. All interdependencies, overlaps, conflicts etc. across projects as well as core and support infrastructure shared across several projects would fall under the purview of this group. Broadly, this capacity at the State Level would:

- Function as a permanent advisory body to the State Government / NeGP State Apex Committee (under chairpersonship of Chief Secretary) in undertaking e-Governance projects.
- Institutionalise processes and tools for effective program management of e-Governance initiatives.
- Provide inputs to the State Government during IT policy/roadmap making process
- Assist individual departments of the State Government in putting together dedicated teams for undertaking e-Governance projects in their departments and managing the process (i.e. building capacity at the department/project level) through the constitution of individual project groups
- The exact nature of support provided by the SeMT to the program would vary over a period of time. Initially it would be to assist the State Government in providing direction & prioritisation and subsequently it would evolve to a program management role.

PeMT: This capacity building is at the department level and provides support to the departments for preparation of the DPRs, business process re-engineering, change management, financial sustainability planning, getting technology expertise and managing the implementation of the project. The objectives of PeMT include:

— A dedicated capacity is required by the Departments in the State Government for conceptualisation and management of the individual projects. On a broad level the PeMT at the Department/project level would:
— Work in close consultation and co-ordination with the departments to conceptualise and prepare Project Proposals
— Identify (and plan the process) for BPR (administrative, legal and process changes) required for improving the efficiency of the department through e-Governance
— Provide assistance to the department for change management, financial sustainability, technology expertise
— Provide project management and implementation support
— Assist the Department in institutionalising the processes and tools required for managing the program
— Engage agencies for hardware, networks, software development, data entry, etc. during the implementation of the projects

Areas

The areas identified in all the tiers of Governance, which will require some specific skill sets for implementing e-governance projects are:

1. Financial management, for
    — Ensuring long term financial viability and sustainability of a project
    — Institutionalisation of mechanism for meeting operational costs
    — Ensuring efficiency and efficacy in resource allocation against various expenditure categories
    — Conceptualisation and development of Public Private Partnership (PPP) models
    — Ensuring inventory (or asset) tracking
    — Ensuring quality in financial reporting

2. Change Management and Process Reengineering, for

   — Creating suitable plans to deal with organizational issues that arise during implementation of e-governance projects. E-governance projects tend to redefine power equations within departments. Stakeholders outside government are also often impacted. Some people lose power, some people gain. This leads to stresses and strains within and outside the organization, which if not managed with finesse, lead to early demise of the project. The first challenge is to clearly identify all internal and external stakeholders in a project, their goals, the extent to which these can be harmonized and factored in, the likely impact on all stakeholders and their likely reaction. This forms the basis of a credible and effective change management plan

   — Ensuring continuity of the program and project in the event of change in key persons / project champions, during the course of the project

   — Synchronization of individual initiatives with the State level / National level initiatives

   — Providing "capacity ramp-up" plan for manpower scaling

   — Providing full ownership to the State Government and Government Departments for implementing the projects

   — Catalyzing administrative, process and legal changes required within the State Government

   — Providing stronger focus on change management through efficient communication, workshops, etc.

   — Alignment of interests of the key stakeholders (Central Line Ministries, State Governments, Department of IT at Centre and State, etc.) thereby synergising various efforts

3. Technology leadership, for

   — Ensuring quick turnaround in developing / conceptualising the technological solution for each department / functional area which is interoperable, standardised and scalable; taking it to a point where implementation agencies could be deployed for specific work.

   — Development of standards for systems, applications and processes across the State

- Development, integration and leveraging of infrastructure of various departments across geographies for reducing overall investments

4. Program/project management, for
   - Enabling administrative and legal changes through involvement of top leadership in the state
   - Adoption of a project oriented approach for the e-Governance initiatives in the State
   - Development of program and project plans integrating timelines, roles and responsibilities
   - Monitoring, tracking and reporting status regularly
   - Development of a communication plan
   - Establishment of project control mechanisms like responsibility matrix, escalation matrix, etc.
   - Undertaking mid-course corrections (if required)

Capacity Building amongst citizens

It is important to understand that even if e-Government initiatives are implemented successfully, it would still be a big challenge to ensure that citizens use the services effectively. Though India has emerged as a major player in the sphere of IT enabled services and the knowledge industry, the movement has, by and large, remained urban oriented, with the exception of a few rural initiatives in some States. The urgent requirement today is to address the digital divide in the country in an integrated and holistic manner. This is only possible when the citizens are made aware about the e-Governance initiatives and how they can effectively utilise it in their day-to-day affairs.

The Capacity Building amongst citizens could be done through mobilization of all relevant tools like Internet, cable TV, community/FM radio and the vernacular press. Combined with appropriate content, connectivity and Capacity Building measures the media can help in usher in higher awareness. Since India has opted for a model of assisted access, particularly in rural areas, building capacity amongst the service centre operators is a key area of attention. Educating citizens about their powers under the Right to Information Act is another dimension. It is now recognized that e-governance has the potential to catapult India on the development path and to facilitate better quality public services in areas such as health, education, environment, provision of business services providing market prices and other information, e-trade opportunities, e-entertainment, e-banking, e-learning, digital photos, e-booking of tickets, internet linked services, etc.

References

Garson, D.G. (2006). Public Information Technology and E-Governance. Sudbury, MA: Jones and Bartlett Publishers.

Rossel, Pierre, and Matthias Finger. "Conceptualizing e-Governance." Management (2007): 399-407.

Saugata,B., and Masud,R,R.(2007. Implementing E-Governance Using OECD Model(Modified) and Gartner Model (Modified) Upon Agriculture of Bangladesh. IEEE. 1-4244-1551-9/07.

4

Promise of E-Government in Developing Countries

The aim of this chapter is to illuminate the evolution, progress and promise of e-government in developing countries and beyond. Beyond in the sense that it will cover the issues, which would be affecting not just the developing countries but also all countries, wherever possible the paper will look into the experience and examples in developed countries and will draw up analogy. The possibilities are exciting: faster access to government services, lower costs for administrative services, greater public access to budgets and documents and a corresponding increase in the transparency and accountability of government activities. However, with the promise of e-government there would also be the challenges. For instance, the increasing threat of 'Digital Divide' causing wider difference between the haves and the have-nots. Where the Internet is not available, the technological 'expertise' untapped and incentives lacking, the concept of e-government seems highly impractical.

It is not difficult for people in developed countries to imagine a situation in which all interaction with government could be done through one counter 24 hours a day, 7 days a week, without waiting in lines. However, to achieve this same level of efficiency and flexibility for developing countries is going to be far more difficult. Experience in developed countries show that this would be possible if governments are willing to decentralize responsibilities and processes, and if they start to use electronic means. Each citizen could then contact the government through a website where all forms, legislation, news and other information made available. It is assumed that the governments of developing countries, as a collector and source of information, could follow this trend, to serve its customers (citizens and businesses) better and to save costs by making internal operations more efficient, cutting down the complex and over stretched bureaucratic system.

There are many definitions of e-government, and the term itself is not universally used. Definitions of e-government range from 'the use of information technology to free movement of information to overcome the physical bounds of traditional paper and physical based systems' to 'the use of technology to enhance the access to and delivery of government services to benefit citizens, business partners and employees'. The common theme behind these definitions is that e-government involves the automation or computerization of existing paper-based procedures that will prompt new styles of leadership, new ways of debating and deciding strategies, new ways of transacting business, new ways of listening to citizens and communities, and new ways of organizing and delivering infor-mation. Ultimately, e-government aims to enhance access to and delivery of government services to benefit citizens.

More important, it aims to help strengthen government's drive toward effective governance and increased transparency to better manage a country's social and economic resources for development. The differences are not just semantic and definitions and terms adopted by individual countries have shifted, as priorities have changed, and as progress has been made towards particular objectives. In the context of the OECD e-Government Project, the term 'e-government' is defined as: 'the use of information and communication technologies and particularly the Internet, as a tool to achieve better government'.

E-governance is more than just a government website on the Internet. The strategic objective of e-governance is to support and simplify governance for all parties; government, citizens and businesses. The use of ICTs can connect all three parties and support processes and activities. In other words, in e-governance electronic means support and stimulate good governance. Therefore, the objectives of e-governance are similar to the objectives of good governance. Good governance can be seen as an exercise of economic, political, and administrative authority to better manage affairs of a country at all levels.

However, as regards to the objective of e-government a distinction should be made between the objectives for internally focused processes (operations) and objectives for externally focused services. The external objective of e-government is to fulfil the public's needs and expectations satisfactory on the front-office side, by simplifying the interaction with various online services.

The use of ICT in government operations facilitates speedy, transparent, accountable, efficient and effective interaction with the public, citizens, business and other agencies. In the back-office, the internal objective of e-government in government operations is to facilitate a speedy, transparent, accountable, efficient and effective process for performing government administration activities. Significant cost savings (per transaction) in government operations could be the result.

Identifying potential gains from e-government is one thing; actually realising them is another. Implementing ICT projects, particularly large-scale projects that could have a major impact on service quality improvements or efficiencies, could raise a number of problems, many of which relate particularly to operating within government.

In order to examine the risk of implementing e-governance solutions the following factors have to be taken into account:

- Political stability (democracy or dictatorial regime)
- Adequate legal frame work
- Level of trust in government (perception of service levels)
- The importance of government identity (fragmentation or integration)
- Economic structure (education, agriculture, industry or service)
- Government structure (centralized or decentralized)
- Different levels of maturity (weakest part of the chain determines speed)
- Constituent demand (push or pull)

These external e-government barriers often concern breakdowns, missing components or lack of flexibility in the government wide frameworks that enable e-government. To adapt to the e-environment, governments need to establish a legal framework that treats electronic processes and traditional processes equally.

Evolution of E-government

E-government refers to the use by government agencies of information technologies that have the ability to transform relations with citizens, businesses and other arms of government. These technologies could serve a variety of different ends: better delivery of government services to citizens, improved interactions with business and industry, citizen empowerment through access to information, or more efficient government management. The resulting benefits can be less corruption, increased transparency, greater convenience, revenue growth, and/or cost reductions. Traditionally, the interaction between a citizen or business and a government agency took place in a government office.

With emerging information and communication technologies, it is possible to locate service centres closer to the clients. Such centres may consist of an unattended kiosk in the government agency, a service kiosk located close to the client, or the use of a personal computer in the home or office. The e-government is supported by the regulations and policies of e-governance, bringing together IT governance and global governance, as well. This is in turn supported by the digital society developed by the connected environment. However, the society must contend with the effects of the digital divide, which represents those members of the society who cannot or will not access the connecting infrastructure.

Caldrow wrote in 1999 'the term e-government has attained conversational status in just four short years'. Therefore, even in the USA, a country that is the pioneer in innovation processes, the term has been circulating for just seven years. On a European level, instead, it was in 1999 that the foundations were laid for 'Europe 2002, an Information Society for All' the first action plan of the European Union for the Society of Information, which is now being substituted by 'Europe 2005; an Information Society for All'.

What does it take to become an 'e-government'? To start with, the answer is not being simple. The fact is that 'the quest is multi-dimensional across leadership, policy, economic competitiveness, education, digital citizen services, internal government operations, digital democracy, and enabling technologies for each dimension'. Consequently, the first characteristic is according to IBM, the multidimensionality of the e-government process. In order to construct a lasting competitive advantage this multidimensional factor needs to be implemented through the 'seven e-government leadership milestones': 'integration, economic development, e-democracy, e-communities, intergovernmental, policy environment, next generation Internet'. In IBM's vision, e-government is the foundation framework within which to operate all the different changes in the different sectors, for which it is unthinkable for example to implement e-democracy without e-government. It is now clear how the latter does indeed represent, effectively speaking, 'a change of mind', a total shift in perspective as utilized in the past, a revolution in the way of being of the public administration towards its interlocutors.

Evolution of e-government has some degree of similarities with the evolution of e-com-merce. Analogous to e-commerce, which allows businesses to transact with each other more efficiently (B2B) and brings customers closer to businesses (B2C), e-government aims to make the interaction between government and citizens (G2C), government and business enterprises (G2B), and inter-agency relationships (G2G) more friendly, convenient, transparent and inexpensive. Tapscott identified such an evolution sequence for e-commerce. He suggested that organizations matured from initially having a web presence only, to offering transaction services and business front-end and back-end integration (as well as several more sophisticated development levels).

E-government also matures along a similar development path, which starts with broadcasting, then interaction, followed by transaction, and finally integration. In broadcasting mode, the government's presence would be made with static web pages and one-way communication. The format of the early government websites is similar to that of a brochure or leaflet. The value to the public is that government information is publicly accessible; processes are described and become more transparent, which improves democracy and service. Internally (G2G), the government can also disseminate static information with electronic means.

In interaction mode, websites would be able to exchange information or services with citizens, where citizens can enquire, and obtain resources from database backed websites located behind a portal. The interaction between government and the public (G2C and G2B) is stimulated with various applications. People can ask questions via e-mail, use search engines, and download forms and documents. The fact is complete intake of (simple) applications can be done online 24 hours per day. Normally this would only have been possible at a counter during opening hours. Internally (G2G), government organizations use LANs, intranets and e-mail to communicate and exchange data.

At the transaction stage, the public would be able to carry out (financial) transactions with the government. This would require higher levels of processing capability, as well as payment gateways and security implementation. In this situation, complete transactions could be done without going to an office. Examples of online services are filing income tax, filing property tax, extending/renewal of licenses, visa and passports and online voting. Phase three is made complex because of security and personalization issues. E.g. digital (electronic) signatures will be necessary to enable legal transfer of services. On the business side, the government is starting with e-procurement applications.

Finally, e-governments would reach integration stage where departments collaborate in significant ways to avoid duplication of efforts, and a one-stop contact point is created, which is capable of handling procedures of all involved departments. In this phase when all information systems are integrated and the public can get G2C and G2B services at one (virtual) counter. One single point of contact for all services would be the ultimate goal. The complex aspect in reaching this goal would be mainly on the internal side, eg the necessity to drastically change culture, processes and responsibilities within the government institution (G2G). Government employees in different departments have to work together in a smooth and seamless way. In this phase cost savings, efficiency and customer satisfaction are reaching highest possible levels. Hence, the forces that once converged to promote e-commerce are now driving us to e-government, along with that, the computer literate population in developed countries are increasingly accustomed to self-service. Changes in the population and specifically in the workforce are forcing governments to 'do more with less'. The trend in the developed countries is to spend a record budget surplus in funding its e-government expenditures.

Because every society has different needs and priorities, there is no one model for e-government and no universal standard for e-government readiness. Each society's and government's readiness for e-government would depend upon which objectives and specific sectors it chooses as priorities, as well as the resources available at a given point in time (which might depend on budgets, donors, etc). The necessary pre-conditions for e-govern-ment depend upon a society's most important needs. For example, the level of infrastructure, legal framework and human capital needed for e-government vary with the objectives being pursued. It is also politically intriguing the idea of implementing a generic and adaptable model e-government as it clashes with the understanding that each country and region has its own peculiarities, constitution, and legal and political framework. The idea is also simply unrealistic.

Once a vision and priority sectors for e-government are established, it is important to assess how prepared a society is for e-government. Assessing e-government readiness requires examination of government itself institutional frameworks, human resources (including ICT managers, procurement officers, and others), existing budgetary resources, inter-department communication flows, etc. National infrastructure, economic health, education, information policies, private sector development and other issues are also factors of society's readiness. Even in developing countries where problems of low connectivity and human resource development (including low ICT literacy) are severe, creativity and careful planning can develop specific applications, services and information that can be delivered in a targeted, useful way to identifiable audiences. A study by Anderson Consulting found vast differences among countries in the maturity of their e-government effort. Perhaps the key finding, however, was that even the most mature countries have tapped less than 20% of the potential.

E-GOVERNMENT AND DEVELOPING COUNTRIES

One of the success stories of e-government is 'eEurope'. The eEurope initiatives have been both highly politicized and highly results oriented. The EU has showed its ability to track, monitor and evaluate the implementation of each Action Plan, using a comprehensive system of key indicators and benchmarking reports produced and submitted to leaders for review and further action. The deliberately detailed implementation of eEurope and the political commitment of the EU leaders to construct and actually enforce a knowledge-based economy and society 'for the benefit of all', distinguish the EU from other global, regional and sub-regional actors. The positive effects of political will and enforcement through effective implementation benchmarking are perhaps the first lessons that developing countries may well learn from eEurope.

What is more important, ICT or clean water? No country, be it developed or developing, large or small can afford to ignore those gains that ICTs potentially offer. For developing countries this is even more pressing, given the 'leapfrogging' opportunities available through modern technologies. I will discuss 'leapfrogging' later in this section. ICT and e-governance, coupled with smart and timely government policies, have the potential to reach development objectives faster and at a lower cost than conventional approaches. One particularly strong opinion questions the very relevance of information society in circumstances of persisting poverty, lack of clean drinking water, widespread corruption, and other characteristics of developing countries that require handling immediately. One cannot deny the vital importance of these and other problems, however it is not possible to avoid that which is unavoidable. The delay in implementation would only cause added cost. However, one can emphasize the importance of regional and sub-regional cooperation in promoting the philosophy of the information society.

E-government for developing countries raises some hopes and demonstrates opportunities. Hence, where, do the developing countries stand in respect of the stages of evolution. Before going into the details about the legal implications, it is essential to trace here the required infrastructure of the developing countries in this context. The objective here is to understand the balance of revenue spent to develop such capability and level of utilization. It is beyond any argument that in the context of developing countries, ICT is one of the most significant forces of modernization. Investment in communication infrastructures, and in training and learning, would enable developing countries to increase their competitiveness as external service provider.

A 2001 study of the development of e-Government by UN identified the level of e-government in 190 nations. The study outlined five stages of e-government, spanning from emergence to integration. At the time of the survey, none of the surveyed nations had achieved integration, and only 17 had achieved the transaction stage. Most developing nations were either at the emergence or at the broadcast stage, thus providing very few interactive services to their citizens. Nations with extensive interactive, knowledge supported services typically also maintained a considerable IT and government infrastructure to offer these, and were generally well funded. This raises the question whether e-Government is largely a question of resources, and specifically whether knowledge management requires the provision of considerable financial and IT resources.

The feasibility of having a successful e-government is directly depended on the governments' overall ability and readiness to spend on the necessary information technology and related costs. Many governments choose not to separate these two and now consider all government IT spending to be e-government spending. The table clearly shows the vast spending differences between developed and developing economies. However, there is another issue to look into. Conventional budgetary structures (such as the one followed by most of the OECD countries: vertical funding structures, in accordance with the core public management principle of holding an agency accountable for achieving organizational objectives and giving it the resources to accomplish those objectives) may not take into account the specific needs of certain most e-government projects, particularly those involving long-term funding requirements and collaboration across agencies. In order to maximize the benefits of e-government for developing countries financing issues must be addressed.

Development of e-governments is directly proportional to the IT infrastructure that is capable of supporting and enabling the execution of e-government. An e-government infrastructure in general comprises network infrastructure, security infrastructure, application server environment, data and content management tools, application development tools, hardware and operating systems, and systems management platform. However, many developing countries do not have the infrastructure necessary to deploy e-government services throughout their territory. A recent study by Harris compared the ICT situations in six selected developing countries, based on World Bank data for 2000. Table 6 illustrates the availability of two important technologies, telephone lines and personal computer penetration, for several developing countries, as well as the number of Internet users. An interesting finding is that the developing countries are very non-homogeneous with respect to technology availability and use, differing by two orders of magnitude.

In a developing country, the gap between the educated elite and uneducated poor is wide. The educated population have the necessary resources and have the means to use information and communication technologies. There is the possibility that professionals in large cities in China or India may find it about as easy to access computers as their counterparts in the UK or Canada; however the situation would be clearly different for those living in rural areas. This 'digital divide' between those who already have access and those who would not gain access for a long time may result in long-lasting and widening economic gaps between the ICT haves and have-nots. As a result, the provision of e-government services would be biased, favouring educated, urban residents.

The potential digital divide is clearly a concern, and yet it applies to many types of public infrastructure, be it roads, airports, or schools. In fact, ICT may offer more hope in this respect than other infrastructures, as technology and communication costs are continuously dropping, and energy for computers can be provided through small power generators or solar cells. As experiences shows, a local village may not have roads, a schoolhouse, or telephone landlines, but with a satellite dish and a cyber cafe, it is possible to be connected to the world. However, connecting people by making investments in the technology would be comparatively an easier job than bringing and keeping them online. The key to encourage citizens in developing countries to use the Internet would be to provide them with compelling content and services that meet their primary needs.

Most of the developing countries understand the enormous potential of ICT, not only as a tool for improving governance and creating more jobs, but also more significantly as a means to enhance the standard of living of the people. The ICT policy aims at increased application of IT in all occupations, enhancing the IT industry base, creating a robust state information infrastructure and creating human resources for IT. Although the policy statements of differs from country to country however there are some fundamental similarities in the objectives, which can be summarized as following:

- Upgrading of the standard of living of the people of the state through use of IT in all sectors as a tool to enhance productivity, efficiency and optimum utilization of resources, and through full exploitation of the employment potential of the IT sector.
- Establishment of an information infrastructure comprising a high speed broadband communication backbone, nodes, access network, distributed data warehouses and service locations to cater to the needs of trade, commerce, industry and tourism and also to enhance the delivery of government services to the people.

- Facilitating the flow of direct investments.
- Development of human resources for ICT through increased use of ICT in educational institutions and through academic and training programmers that improve the em-ployability of educated youths in the ICT sector.
- Facilitation of decentralized administration and empowerment of people through the application of ICT.

Currently, there are insufficient numbers of people in developing countries trained in appropriate technologies available to do all the work. Training opportunities are also straining to meet needs. The digital divide, while closing, demands that multiple-delivery channels be maintained for the near future. Many people are concerned about the issues surrounding privacy and confidentiality, even to the point of opting out of e-government opportunities. Technologically, the lack of a single, unifying infrastructure coupled with differences in implications, applications and data structures increase the difficulty of sharing information between departments and levels. Labour rules and policies particularly in public sector can strongly influence how new tasks and skills are introduced. Geographical boundaries, less important to commerce and communication, are still ingrained in the political process. In some cases, the use of the Internet or other channels has been mandated for government use, whether or not doing so makes sense technologically. Although the expanding economy is providing the funding to fuel the necessary expenditures for e-government, the regulatory framework and funding models are lagging, a situation further complicated by a model in which the groups paying for services are not necessarily the recipients of the services. Finally, although government services are typically seen as environmentally clean, the impact of providing high bandwidth to the curb can be enormous.

However focusing on the achievements of e-governance and ICT for developing countries it would be prudent to mention here some of the success that ICTs presented particularly the opportunities to the disadvantaged to help themselves, seize advantages and demand services or control over their lives that they previously lacked. These examples are multiplying, as collaborative approaches and partnerships between the public and private sectors are demonstrating the broader value of advanced communications services:

- In Brazil's urban slums, the Committee to Democratise Information Technology (CDI) in four years has created 110 sustainable and self-managed community-based 'Computer Science and Citizenship Schools', using recycled technology, volunteer assist ance, and very limited funds. CDI schools train more than 25,000 young students per year in ICT skills that give them better opportunities for jobs and education. CDI also provides social education on human rights, non-violence, environmental issues, health, and sexuality.
- The Association for Progressive Communications and Partner Networks implemented a Gender and Information program. The program gave women in Africa, Asia and Latin America the necessary technical skills, computers, hardware, and networking to plan the UN's Fourth World Conference on Women and to disseminate the results.
- An example of how new networking can help sub-national governments to become more responsive and transparent is the case of the state of Andhra Pradesh in India. Andhra Pradesh was the first state in India to design a statewide computerization program covering all levels of the administrative spectrum from the smallest—the Mandal Revenue Offices—to the largest (eSEVA). The state-wide Multi-Purpose Household Survey was launched in 1995 to store large volumes of citizen data and information exchange through networking, and Land registration offices throughout the state are now equipped with computerized counters under the Computer-aided Administration of Registration Department (CARD) project. Before the computerization programme, tedious bureaucracy, and lack of transparency in property valuation resulted in a flourishing business of brokers and middlemen who exploited citizens selling property, since a premium was usually attached to speedy delivery of services.

The success of these projects has reaffirmed the faith in what has been termed as 'leapfrog' technology. New, innovative and cutting edge technology could help developing countries overcome their traditional barriers of lack of telecommunications infrastructures. In developed economies, newer versions of technology are often used to upgrade older versions, but in developing economies where still older versions of technology are often prevalent, the leapfrogging over the successive generations of technology to the most recent version creates more possibility and often provide access to newer, efficient and often cheaper alternatives to traditional capital intensive technologies.

As the situation stands today it is in the interest of the developed countries to play an important role in promoting technology adoption in developing countries, with international organizations such as UN or the World Bank providing the necessary funds. However, pushing e-government initiatives to achieve a significant impact on establishing a digital society and a connected economy would prove to be an elusive goal without proper investment in basic services such as healthcare, education. E-government solutions and approaches used in developed countries may also prove inadequate without a clear understanding of local constraints and conditions.

E-government and Legal Issues

The success of e-government initiatives and processes are highly dependent on government's role in ensuring a proper legal framework for their operation. A requirement for e-government processes to be introduced and adopted is their formal legal equivalence and standing with the paper process. OECD governments are aware of the need for framework to provide for enforceable electronic transactions, both in the e-government sphere and for e-commerce, and have taken action. For example, the legal recognition of digital signatures is necessary if they are to be use in e-government for the submission of electronic forms containing sensitive personal or financial information. As of 2003, 26 of 30 OECD countries have passed legislation recognising digital signatures, though a much smaller number have actually introduced applications beyond a pilot phase.

The key concern for the developing countries would be to identify the legal issues. As explained before in this chapter, it would not be difficult to provide definitive guidelines for the developing countries as to what type legal safeguards would be necessary to protect the interests of the government and to create binding and enforceable obligations on the government. However, the developing countries are in different stages or phases of e-government, they differ in their political aspirations and in their structure of government and so their requirements would differ and so would their course of action.

The developing countries would face a few key questions in converting to or adopting electronic processes. To start with, the first question that would arise is the question of legitimacy. Legitimacy may express itself through expressions of authority to act, which is related to but not the same as saying that an action is not illegal. In some settings, the legitimacy of government action would affect the enforceability of the action, either on the part of government against a person subject to that government's rule or on the part of private citizen against the government. Hence, to provide legitimacy to act electronically can only be conferred by legislative measures and supported by an effective legal framework. This framework should be capable of identifying and addressing legal obstacles to e-government. Legal obstacles may include the differences that exist between traditional data collection requirements and the ease of electronically collecting and sharing data.

Legitimacy also turns on the standards of appropriateness. What kind of qualities should e-government action have, or some standards it should meet, in order to achieve the desired level of legitimacy. In certain situations, an electronic process would prove to be 'good enough' to meet its legal needs without regard to whether it is comparable to or as good as its prior process. At the other extreme, would be the situation where electronic conversion would require a complete re-engineering of their course of action in order to address the legal risks and issues that a particular system presents or that are not being addressed as effectively in their existing system. Because most of the developing countries owing to lack of resources would still depend largely on their existing paper-based systems as a reference point in their analysis, hence the legal framework would require to have a point of compatibility, which I think would be quite unique for these countries. As the situation stands, at least in the near future most of these developing countries would require having both systems of transactions. In order to achieve the legislative revisions, enactment would be required, as current laws, rules, and regulations might not recognize the legality of electronic documents and processes. For example, legislation should ensure that electronic authorizations, contracts and signatures would have the same legal effect as those done on paper. Additionally, governments may consider the required quorum at meetings be met through the electronic participation of officials rather than by the officials' physical presence.

It can be said that much of what a government, or those who deal with a government, would look for in an electronic communication would be a latent degree of assurance as to the security of the communications and their source: who sent them? Any user would primarily be concerned about the integrity of the information communicated, in the sense that it is trustworthy and that it has not been altered since it was sent. Similarly, the concern would also be about jurisdiction. However, the issues relating to jurisdiction would be similar in case of both developed and developing countries. The only point of concern is how it should be addressed. E-government for developing countries would also put the judicial system of these countries to test. Generally, the judiciary in these countries are not always in tune with the use and advancement of technology.

Electronic communications appear particularly susceptible to attacks on privacy; in most of the developing countries, privacy is a softer issue and hence even with proper legal framework it can be argued how far privacy of personal information would be maintained. Likewise, fair access to public services is one element that would bring back the issue of 'digital divide' within the developing countries-providing facility to the more able citizens. At some point, this would cause the law to impose constitutional limitations on government's ability to move wholly online in order to maintain equality.

The legal status of government documents is always of high importance simply because of the weight of authority. One sees this in the priority they are frequently given in evidence statutes. This could be considered as a higher form of authenticity accorded to these records, but it has an impact on the demands made of electronic communications so that they would deserve the same respect that paper documents receive. Pushing this theme further, it can be argued that the integrity that the public expects of government communications is not just that they are unaltered but their content is true. That is not a question of the medium used.

Finally, government use of electronic communications has to be politically acceptable to the opinion makers of a society. No amount of technical excellence can guarantee that any manifestation of e-government would meet this standard. One does hear from politicians, and sometimes from business people, that government should lead the way to making people confident about electronic communications being safe and effective. This is a technical aspect of the legitimacy argument that proper government use can encourage general use of paperless records, but how far the governments in the developing countries particularly the politicians themselves are comfortable with the technology is matter of debate.

Not all these concerns are specifically related to developing countries nor do they give rise to legal consequences, but any discussion about e-government for developing countries could not be made without discussing these issues. However, the difference would be in the level of priorities. Some factors would play more important roles in some governments than in the others.

Ideally, a legal framework that allows for the implementation of e-government processes and services will:

- Preserve basic public policy goals, such as privacy and security, retention, and public access to information.
- Provide the statutory basis of, authority for, and regulations related to the government processes and services that may be supplied electronically.
- Assign responsibility for and ownership rights to the data provided and accumulated electronically.
- Address the sharing of data collected by one government agency with other government agencies that require the same information.
- Clearly, define jurisdictional responsibilities related to intergovernmental transactions and business to government transactions.
- Provide a mechanism by which legal requirements are recognized and enforced.
- Provide a basis for the establishment of fees related to electronic processes and services.
- Identify the records that should be maintained, the period of retention and the required storage media.
- Not be technology-specific or favour one form of service delivery (traditional or electronic).
- Minimize costs and the potential for litigation.

Legislative Approaches

As in every other developed country, the traditionally drawn legislation in developing countries, drafted largely in a paper-based environment is formulated in a manner not tailored to the legal recognition of electronically generated, stored and transmitted information. This feature of legal systems, is epitomized by the traditional basic requirements scattered throughout legislation, calling for documents (including, for example, contracts) to be in 'writing', for an 'original' to be submitted to various governmental agencies, and for documents to be 'signed' by the purported author. Such traditional formulations can be found throughout a legislation (e.g. Civil Code, Law of Evidence, Arbitration Law), affecting diverse activities such as filing of documents regarding movement of cargo through ports, admissibility in court of computerized banking records, securities trading and many others. Such paper-based formulations define requirements in public law as to the form of information that has to be provided to governmental authorities by business and citizens for various purposes (e.g. applications for permits and licenses, documents required to be filed with customs, tax and other authorities). Such paper-based notions also apply to the legal validity of various types of communications exchanged amongst the private sector in its commercial activity. Moreover, this legal condition leaves the evidential value in court of electronically generated, stored or transmitted evidence uncertain at best, or renders electronic evidence inadmissible at worst.

Some of the developing countries have introduced legislation following UNCITRAL Model Law on Electronic Commerce. The advantage, indeed the indispensability of utilizing a legislative approach stems from the fact that it is beyond the contractual power of parties to overcome obstacles such as those described above, to the extent that those obstacles are embodied in mandatory rules of law. Moreover, utilizing legislation to address such fundamental issues provides greater legal certainty than reliance purely on contractual solutions. Hong Kong enacted the Electronic Transactions Ordinance (effective 7 April 2000; enacted 7 January 2000), which covers electronic and digital signatures and electronic records. This act is generally applicable to all communications. South Korea's Basic Law on Electronic Commerce also covers digital signatures and is generally applicable to all communications. Malaysia has its Digital Signature Bill of 1997, which became effective on 1 October 1998. Singapore's Electronic Transactions Act of 1998 (enacted 29 June 1998) covers digital and electronic signatures as well as electronic records, and is generally applicable to all communications. Similarly, Thailand's Electronic Commerce Law 2000 covered electronic signatures and is generally applicable to all communications. In the Philippines, the Electronic Commerce Act of 2000 (enacted 14 June 2000) encompasses electronic signatures, electronic transactions and crimes related to e-commerce. The Electronic Transactions Order of Brunei (enacted November 2000) covers electronic contracts, as well as digital and electronic signatures. India's Information Technology Act of 2000 (implemented in October 2000) covers digital signatures and electronic records, and is generally applicable to all communications.

Privacy

The right to privacy is fundamental to any democratic society. It is not just 'a' regulation but it is recognized as a fundamental right in all major international treaties and agreements on human rights and in the constitutions of most countries in the world, either explicitly or implicitly. The slightest apprehension on the part of a person using the Internet about who will see their personal information and how it will be used would by itself mean that they have lost a basic freedom. Moreover, the more others know about the details of a person's life, the greater their opportunity to influence, interfere with, or judge the choices the person makes. Having knowledge and control of how personal information is provided, transmitted and used is the key to protecting privacy.

Ensuring that e-government initiatives are in step with society's expectations in this area is a crucial means of building trust. The challenge facing e-government coordinators and implementers is to respect accepted privacy principles while allowing the benefits of the Internet and other technologies flow to citizens. This balance is of particular importance when considering seamless government services involving data sharing among agencies. Of utmost importance is information privacy, 'individual's claim to control the terms under which personal information—information identifiable to the individual—is acquired, disclosed and used'. The role of governments regarding privacy protection is twofold and therefore always ambiguous. Governments have to ensure the respect of the rule of law, but do also play a proactive role in the development of a welfare state. Finding a correct balance between these two roles of the modern State will always remain one of the core problems in the development of a regulatory framework in this area.

Government has a responsibility to provide leadership in developing a culture of privacy protection and security. It should provide this leadership through its roles in the development of public policy, as owner and operator of systems and networks, and as user of such systems and networks. As a user of information systems and networks, government shares a role with businesses; other organizations and individuals for ensuring secure use of the system and network.

The OECD was the first inter-governmental organization to issue guidelines on international policy for protection of privacy in computerized data processing. In 1980, the Guidelines on the Protection of Privacy and Transborder Flows of Personal Data (Privacy Guidelines) were adopted as a Recommendation of the OECD Council. They were followed by the 1985 Declaration on the Transborder Data Flows, and more recently by the Ministerial Declaration on the Protection of Privacy on Global Networks, adopted by OECD Ministers at the 1998 Ottawa conference, 'A Borderless World: Realising the Potential of Global Electronic Commerce'. At that conference, OECD Ministers reaffirmed 'their commitment to the protection of privacy on global networks in order to ensure the respect of important rights, build confidence in global networks, and to prevent unnecessary restrictions on transborder flows of personal data'.

Among the perceived benefits of electronic transactions between citizens and governments is the opportunity to collect 'customer' information that can be used as a key input to operational and policy decision-making. It is relevant to point out here one such operation where government uses substantial personal information. It is in the case of billing. Billing transactions are a fruitful source of information that allow strategic planners to sift and analyse transactions to identify key trends and unmet service needs, and, because of this, determine the future operational priorities of an agency. For the private sector, that information is an increasingly significant source of business advantage. The variety and sophistication of online identification technologies have increased accordingly, allowing the ready collection and display of an individual's personal details, purchasing history and commercial value. In marketing terms, it allows a focus beyond groups and subsets of groups, to the 'customer of one'. As attractive as this focus might be to public sector planners, the possibility that an electronic transaction will provide an instant window on a citizen's details represents a real threat to personal privacy while online. If the objective of electronic billing is to encourage citizens to adopt a low-cost transaction channel, the belief that it will permit unwanted intrusions by government agencies into personal information is likely to act as a major disincentive. Technology also further complicated the issue by what is known as 'Caching' and 'Cookies'. 'Caching' occurs when any web page accessed by a user is stored by that client's computer (client caching) or by the network server that provides the user with access to the internet (proxy caching). Caching causes a potential problem where it occurs for profiling purposes. The use of 'cookies' is more intrusive, as they have the capacity to build profile on the needs, preferences and patterns of expenditure of any individual visiting particular web sites. Cookies work by placing an identifying code on the hard drives of those who visit the site. This code allows the visitor to be tracked as they travel through the website and to be recognized on subsequent visits.

These technological developments have increased the surveillance potential of the government and prompt the demands for specific rules governing the collection and handling of personal information. The question could no longer be whether information could be obtained, but rather whether it should be obtained and, where it has been obtained, how it should be used. A fundamental assumption underlying the answer to these questions is that if law allows the collection of personal information, the fairness, integrity and effectiveness of such collection and use should also be protected.

Regrettably, most of the legislation mentioned above does not specifically offer a broad protection of individual privacy. In principle, OECD guidelines specify the following for protection of privacy:

- Personal information must be collected for a lawful purpose; Personal information must be collected directly from the individual concerned; The individual must be made aware of a number of matters (including that information is being collected, the purpose for which the information is being collected, the intended recipients of the information, and the name and address of the agency collecting the information);
- Personal information must not be collected in an unlawful or unfair manner; Personal information must be protected by adequate security systems; Individuals are entitled to have access to, and request correction of, personal information held about them;
- Personal information shall not be used unless it is accurate, up to date, complete, relevant and not misleading;
- Agencies must not hold personal information for longer than it is required; and Personal information must not be used for any purpose other than that for which it was collected.

However, none of these principles excludes the use of billing transactions to create citizen profiles. They do, however, require that such use is both transparent and done with the consent of the individual concerned. The question for public sector agencies, then, is whether the perceived benefits of this offset the risk that the bulk of citizens will avoid electronic transactions in favour of traditional and therefore less intrusive billing and remittance channels.

Privacy must be addressed in the planning and design of e-government systems since it is much harder to interject privacy protections after a system is built. Government websites and online services should fully comply with the fair information principles outlined. How far the developing countries recognize these issues as legal obstacles is a matter open to argument. Business interests govern policies and practices in most developing countries. Hence governments would find that proposed measures to protect privacy meet the staunch opposition of business interests which see such safeguards as an expense and an unjustified constraint on their right to conduct their business affairs as they wish. Hence, finding a balance between the legitimate need to collect information and the need to protect privacy would become a major challenge for the developing countries.

The developing countries that have developed legislations or are currently developing laws are doing it in an effort to promote e-commerce. These countries recognize that consumers are uneasy with the increased availability of their personal data, particularly with new means of identification and forms of transactions, and therefore that their personal information is being utilized worldwide. Data privacy laws are therefore being introduced, not from a human rights perspective, but rather as part of a package of laws intended to facilitate e-commerce by setting up uniform rules

Authentication

Authentication is an element of security. Security can be divided into two elements: network security and document security. Security is never absolute; it is relative. The legal implication of an insecure network is caution about what one commits to the network, and how one keeps backup copies of documents. The second element of security is document security, which essentially influences how sure one can be of the answers to three questions: 'what?', 'who?' and 'where?'. Document integrity is the first:

has the document been altered from what was intended by the parties to a communication? The practical issue for integrity is how one keeps one's data from being altered inappropriately or from being assed or destroyed by unauthorized persons. Document source is the second element: who made the document and who sent it? This is usually known as the question of authentication. The issue for authentication is how one can be sufficiently sure that one knows who one is dealing with. The challenges faced by the developing countries is not that much different from the developed countries; with only one possible exception that is, developing counties' governments may be more tolerant of the risk.

How does the government know who it is dealing with? To a lesser extent, how do people dealing with the government know that they are really dealing with the government? There are two elements to this function: identification and verification. There is always a tension between authentication and privacy. As stated before governments have records that can be used to identify people reliability, but privacy laws at least in developed countries limit such activities to known programs and information collected in anticipation of those programs. In practice, much of the discussion of authentication turns on the use of signatures precisely the issue of electronic signatures.

It is not easy to classify the existing legislations with respect to electronic authentication because of the many differences that exist. It is possible, however, to sketch the main approaches at a national and international level. Three approaches can be identified: (1) the digital signature approach; (2) the two-prong approach; and (3) the minimalist approach.

The digital signature approach is characterized by its focus on the digital signature technique. Legislation under this category is truly digital signature legislation because it regulates (on the basis of) digital signatures. Legislation under this approach is concerned solely with the (evidentiary) status of the digital signature. The approach has three variants:

1. *Technical variant:* The technical variant amounts to setting the digital signature technique as a technical standard by means of a legal instrument. The technical variant does not deal with legal consequences, although such consequences may implicitly follow from the use of digital signatures in accordance with the law concerned.

2. *Legal variant:* The legal variant of the digital signature approach is found in legislation that specifically regulates digital signatures in order to provide this technique with a legal status similar to that of the hand-written signature. The general purpose of these laws is to provide legal security for the use of digital signatures. Often legislation of this kind also includes the implementation and regulation of a Public Key Infrastructure (PKI).
3. *Organizational variant:* The organizational variant of the digital signature approach neither sets the digital signature as a technical standard nor provides for explicit legal recognition of the digital signature. Instead, it addresses the organization of Certification Authorities (CAs) and the use of digital certificates in connection with digital signature applications. The aim is to promote trust and reliability in electronic transactions by ensuring that CAs is reliable and secure.

The second approach is called 'two-prong' because of its hybrid way of dealing with electronic authentication. In this approach, legislators aim to make their legislation more time-resistant by addressing certain technological requirements and by leaving room for new technological developments. With this approach, legislation sets requirements for electronic authentication methods that will receive a certain minimum legal status (the minimum prong) and assigns greater legal effect to certain electronic authentication techniques (the maximum prong). The technologies given this higher legal status are referred to as 'secure electronic signatures'.

The minimalist approach does not address specific techniques and therefore intends to be technology-neutral. Legislation relates to the functions that signatures may have to fulfil in trade, and the different levels of reliability with respect to the purposes the signatures are used for. Because the main focus of this approach is on the relevant functions of signatures and the ways in which these functions may be translated into technological applications, it is also called the functionalist approach. Within the minimalist approach, the focus on functions of signatures (and writings) can be more or less explicit.

The market is constantly changing and it is unknown what lies ahead with respect to technological developments. Thus, it might be unwise to issue detailed regulations and to determine specific business models, such as the PKI model, since their viability cannot be ascertained. Viewed in this light, the digital signature approach is seriously flawed. Although, the legislators and regulators subscribing to this approach may do so for all the right reasons (legal certainty, trustworthiness with respect to legal matters), the approach is not recommended as such. The same is true, but to a lesser extent, of the two-prong approach, which attempts to skirt the uncertainties by presenting an opening for new technologies aside from setting criteria for certain advanced electronic signatures which at present cover digital signatures. The approach is understandable in the sense that there seems to be a strong inclination to look for clear and trustworthy solutions, while at the same time there is a need to leave room for new solutions. Still, within the two-prong approach legislation often deals with issues and situations (e.g. CAs, liability, qualities that focus mainly on certain techniques) that have not yet been determined. Finally, both the digital signature approach and the two-prong approach are in many instances focused too narrowly on signatures as such and not on formal requirements as a whole.

The minimalist approach taken in the UNCITRAL Model Law offers the most sensible solution to legislators wanting to tackle the problem of formal requirements in their legislation. Under this approach, legal requirements of form are generally dealt with in their entirety. Moreover, the minimalist approach allows for different functions which techniques have to fulfil under national legal systems, while creating room for new techniques and adventitious developments. Recent legislative initiatives recognize the advantages of the minimalist approach and have explicitly taken the UNCITRAL Model Law on Electronic Commerce as an example.

Much of the discussion above put emphasis on differentiating the legal position, which presupposes the presence of paper, and legal position in the electronic age. However, it is interesting to note why one was using paper, is it because it was required by law or is it because of the convenience of paper's qualities in use. In law, a pencilled 'X' may serve as a signature, for example on the will of an illiterate person. However, most people would not accept a cheque with a pencilled 'X', in the signature line. What is legal and what is prudent might be different.

The law applicable to e-government is evolving quickly, on a framework of information and technology and public expectations in quick development. The narrow questions of legal authority of government to use electronic communications are relatively easy, particularly as they are now expressly dealt with in enabling statutes based on UN Model Law on Electronic Commerce. However, developing countries require specific legislations, which would be capable of accommodating their particular needs. The paper has discussed the main legal question relating to privacy and authentication. It is difficult to legislate with broad application on these subjects, for two reasons: first they to some extent contradict each other, second the technology changes so quickly that laws based on one particular hardware, software or configurations of them are likely to be outdated by the time they are enacted. For developing countries, it is best to be cautious. The law cannot require what technology cannot support. The law for developing countries here is likely to be validating rather than normative. When the gap between a flexible application of current law and demands for new technology grows too great, then legitimacy of e-government comes more severely into question. However different developing countries depending upon their priorities would decide when that point would be and would find a way to act when it arrives.

References

Ignace Snellen, (2002). "Electronic Governance: Implications for Citizens, Politicians and Public Servants," *International Review* of *Administrative Sciences* 68, no. 2, June, pp. 189-90.

Jane E. Fountain, (2001). *Building the Virtual State: Information Technology and Institutional Change.* Washington, D.C.: Brookings Institution Press.

Organisation for Economic Co-operation and Development, *The e-Government Imperative,* p. 156.

Wellman, B. (ed.) (1999) *Networks in the Global Village*. Boulder, CO: Westview Press.

5

Worldwide Development of E-Government

Today, powerful new technologies can be used to advance sustainable development for all people across the world while including them in the process. In particular, e-government can be an engine of development for the people. In delivering e-government for the people, public services are designed to be responsive, citizen-centric and socially inclusive. Governments also engage citizens through participatory service delivery processes. The evidence base for the latter is strengthened by recent progress in e-government in a growing number of countries where citizens are both users and co-producers of public services.

The increasing role of e-government in promoting inclusive and participatory development has gone hand-in-hand with the growing demands for transparency and accountability in all regions of the world. E-government has strongly shifted expectations of what governments can and should do, using modern information and communication technologies, to strengthen public service and advance equitable, people-centred development.

Progress in online service delivery continues in most countries around the world. The United Nations E-Government Survey 2012 finds that many have put in place e-government initiatives and information and communication technologies applications for the people to further enhance public sector efficiencies and streamline governance systems to support sustainable development. Among the e-government leaders, innovative technology solutions have gained special recognition as the means to revitalize lagging economic and social sectors.

The overall conclusion that emerges from the 2012 Survey in today's recessionary world climate is that while it is important to continue with service delivery, governments must increasingly begin to rethink in terms of e-government – and e-governance – placing greater emphasis on institutional linkages between and among the tiered government structures in a bid to create synergy for inclusive sustainable development. An important aspect of this approach is to widen the scope of e-government for a transformative role of the government towards cohesive, coordinated, and integrated processes and institutions through which such sustainable development takes place.

National E-Government Development

The United Nations Survey 2012 assessment of progress indicates that e-government is increasingly being viewed among countries in the vanguard as going beyond service delivery towards a framework for a smart, inclusive and sustainable growth for future generations. In countries that follow that trend, a focus on institutional integration coupled with online citizen orientation in public service continues to be dominant. Both in terms of information and services, the citizen is increasingly viewed as 'an active customer of public services' with borrowed private sector concepts being applied to improve public sector governance systems.

A key driver for this approach is the need to achieve efficiency in government at the same time that services are being expanded. Advances in technology, which allow data sharing and efficient streamlining of cross-agency governance systems are forming the back end of integrated portals where citizens find a myriad of relevant information arranged by theme, life cycle or other preferred use. The trend towards personalization of services has gained momentum with more countries tailoring substance and presentation in accord with varied preferences. Multichannel service delivery features were found on several portals in 2012 through which the government conducted business with citizens. Citizen inclusion is also expanding both horizontally and vertically with more governments around the world in 2012 accepting and promoting the need to inform – and involve – the citizen in the public decision making process.

E-government innovation and development can position the public sector as a driver of demand for ICT infrastructure and applications in the broader economy. The effect will be more pronounced in cases where government programmes constitute a significant proportion of a country's GDP and where the regulatory environment is conducive to expansion of ICT manufacturing, software and related services.

E-government programmes can be a catalyst in boosting productivity, thereby speeding up the benefits of newer technologies to the people. In the last few years many countries have employed ICT in areas such as entrepreneurship, innovation, research and development, promoting distance learning, e-health, e-agriculture, e-trade and other fields. Accessing these new technologies for development is being recognized as one of the key sources of economic growth. Of particular importance is the effect of cellular technologies. Where national governments have taken a lead, rapid mobile technology proliferation has contributed as much as a one per cent annual increase in economic growth over the last few years.

Notwithstanding these trends, progress remains uneven. In the current recessionary climate some countries have been bet er able to continue to invest in ICT infrastructure and service improvement. Others are evaluating the marginal utility of such investment, especially taking into account low user uptake of existing services, and reassessing service portfolios where demand for online services is low. Many countries with low levels of infrastructure and human capital remain at lower levels of e-government development with serious issues of digital divide.

In all cases, e-government take a prominent role in shaping development making it more in tune with people's needs and driving the whole process based on their participation.

Global Leaders at a Glance

Building upon the transformative nature of ICT and maintaining their focus on e-government development, all of the top 20 countries in 2012 were high-income developed economies. All have values that range from 164 to 190 per cent of the world average. Of the 20, 14 are in Northern America and Europe; 3 in East Asia (Republic of Korea, Singapore and Japan); 2 in Oceania (Australia and New Zealand); and 1 in Western Asia (Israel).

While the Republic of Korea (0.9283) maintains its position as achieving the greatest e-government development, in 2012 it is followed by three European countries, with the Netherlands (0.9125) advancing by three and the United Kingdom of Great Britain and Northern Ireland (0.8960) by one to become the 2nd and 3rd leading e-ready governments in the world. Denmark (0.8889), the United States of America (0.8687), France (0.8635) and Sweden (0.8599) follow close behind among the global leaders.

In 2012, the United Nations e-government assessment focused on the concept of integrated services that exploit inter-linkages among different public services on a functionally and/or thematically similar one-stop-shop portal, thereby improving and facilitating citizen experience, allowing for back-office integration across governmental departments and strengthening institutional arrangements. Single sign-on integrated services on portals can organizationally transform public service delivery at both the front and the back end. They can increase functional productivity in governments by identifying and improving governance processes and mechanisms across several departments, leading to greater efficiency and effectiveness of services along with needed cost savings. With a focus on governance solutions that enhance service delivery and streamline public sector efficiency, the United Nations e-government rankings in 2012 reflect an assessment of which countries are undertaking their e-government development with a view to integrated, user-centric public service delivery.

The 2012 Survey assesses web portals with a view to the provision of e-information, e-services, which range from interactive to transactional to networked services, e-participation, and features that are the conduit for service flow from government to citizen. Indicators grouped along the four stages of the model (emerging, enhanced, transactional and connected) range from static information such as links to ministries/departments, archived information, and regional/local government services; to unidirectional government-to-citizen (G2C) information flows such as online policies, laws and regulation, reports, newsletters, and downloadable databases, among other things; to two-way financial and non-financial transactional services and advanced technical features such as mobile apps; and to integrated and participatory services characterized by an integration of government-to-government (G2G), government-to-citizen, and citizen-to-government (C2G) interactions in the last stage.

The United Nations Survey 2012 finds that models of an integrated portal differ across countries and regions. While a few countries are progressing towards one national integrated portal, others have developed their e-government offerings with a view to more than one portal, with thematic and/or functional services integrated in a manner that finds e-information separate from e-services or e-participation.

Though each of these have integrated services across various departments on the thematic or functional portal, they nevertheless make less convenient the user search for government information, services and participation in one place. The United Nations E-Government Survey 2012 differentiates these as 'integrated services' from a single 'integrated portal.'

In 2012 no country had a true single-sign-on integrated portal. The United States, Republic of Korea, Israel, Australia, Norway, Denmark, Bahrain, Qatar, United Arab Emirates and New Zealand are among the few that come close to a pure one-stop-shop portal with information, services and participation services integrated on one site.

Most countries from the European Union (EU) follow the approach of separate portals for their information, service and participation offerings. In several European countries e-government services focus on the nationally organized one-stop channel for the provision of 20 basic e-services essential to their citizens while the government-provided information forms a separate portal with information services integrated on it from across all sectors.

Lessons of experience from the assessment in 2012 indicate that more services have been integrated across sectors and agencies. While this trend is likely to continue it seems that increasingly complex public sector services in the future will be 'cloud-based' with service providers able to address innovation and productivity upgrades without costly investments by the government. Cloud service equips governments with greater efficiency by helping them scale up their services, including storage capacity, as it evolves. Among the main challenges for large-scale adoption of cloud-based government services are the integrity of service, data security and privacy, and regulatory environment in most countries around the world, which will need continued reform in governance systems and a continued and a continued focus on strengthening institutional linkages.

Close behind the top world leaders are the 25 emerging leaders. Among these, 16 are in Europe, 6 in Asia and 3 in the Americas. With close proximity in the e-government development index value, the leaders among this group are Austria (0.7840), Iceland (0.7835), Spain (0.7770) and Belgium (0.7718). Substantia l effort was made by some countries, which is reflected in their advancement this year. Notable among these are the

Russian Federation (0.7345), the United Arab Emirates (0.7344), and Saudi Arabia (0.6658), all three of which joined the emerging leaders group. Progress was also noted in the case of Italy (0.7190) and Portugal (0.7165).

Table 1.World e-government development leaders 2012

Rank	*Country*	*E-government development index*
1	Republic of Korea	0.9283
2	Netherlands	0.9125
3	United Kingdom	0.8960
4	Denmark	0.8889
5	United States	0.8687
6	France	0.8635
7	Sweden	0.8599
8	Norway	0.8593
9	Finland	0.8505
10	Singapore	0.8474
11	Canada	0.8430
12	Australia	0.8390
13	New Zealand	0.8381
14	Liechtenstein	0.8264
15	Switzerland	0.8134
16	Israel	0.8100
17	Germany	0.8079
18	Japan	0.8019
19	Luxembourg	0.8014
20	Estonia	0.7987

It is somewhat noteworthy that the emerging leaders group includes some developing countries that have begun to catch up with higher-income countries, such as Kazakhstan (0.6844); Chile (0.6769), Malaysia (0.6703), Colombia (0.6572), Barbados (0.6566) and Cyprus (0.6508).

Many of these countries have invested considerable resources in e-government in the last few years. They have expanded infrastructure and human skills on which to build further advances in service delivery and employ the full potential of information technologies for long-term sustainable development. Some of the developing countries have found ways to leapfrog traditional development cycles by deploying mobile technology for bridging the digital divide. They have reoriented their public sector governance systems towards user-centric approaches visible on their websites through multichannel service delivery features.

As in the case of the world leaders, countries in the emerging leaders group have e-government development values close to each other, ranging from 0.6508 to 0.7840. Most of them are providing similar levels of e-services such as in the case of Cyprus, which though ranked lowest for this group, has achieved around 83 per cent of the level of e-gv-ernment development of Austria, the group leader.

Countries with a Large Population

The raison d'être of the United Nations E-Gov-ernment Survey is to assess whether countries are deploying e-government for inclusion-for-all.

Since each country faces a different set of factors that can help or hinder its overall progress towards e-government development, this year the United Nations Survey is extending special recognition to those countries which, with a population of over 100 million, have made a tremendous effort to provide e-government services to their people, despite the challenges they face.

Table 2 presents e-government development in countries with populations larger than 100 million that have made a special effort to improve service delivery to large swathes of their populations. It should be kept in mind that the E-Government Development Index (EGDI) is constructed on a comparative basis that rates each country relative to all other Member States.

Lack of access to both ICT and education infrastructure in the developing countries is a major constraint on e-government development. Income per capita imposes another limiting factor, with lower income countries having a higher marginal cost for a dollar spent on ICT. With the economic downturn, online services are at a disadvantage in the competition for resources with safe water, rural health and basic education services. This becomes especially acute if the country has a large population and/or a large land area since e-inclusion demands that online service access and infrastructure be available to all. Large areas require greater investments in providing telecommunication infrastructure. Even with cellular technology on the move, connectivity remains a major challenge for far flung rural areas. Including a population of 200 people living in the Sahara desert denotes a high marginal cost for the government. Large populations also require greater investments in schools and functional literacy. Many developing countries continue to feel the drag of a low level of educational achievement, which pulls down the United Nations E-Government rankings.

Table 2. E-government development in largest population countries

E-gov. development index			***World e-gov. development ranking***		
Country	*2012*	*2010*	*2012*	*2010*	*Population (in millions)*
China	0.5359	0.4700	78	72	1,341
India	0.3829	0.3567	125	119	1,225
United States	0.8687	0.8510	5	2	310
Indonesia	0.4949	0.4026	97	109	240
Brazil	0.6167	0.5006	59	61	195
Pakistan	0.2823	0.2755	156	146	174
Nigeria	0.2676	0.2687	162	150	158
Bangladesh	0.2991	0.3028	150	134	149
RussianFed.	0.7345	0.5136	27	59	143
Japan	0.8019	0.7152	18	17	127
Mexico	0.6240	0.5150	55	56	113

Implicit in the concept of inclusion-for-all is that large, low income countries must exert far more effort to achieve a given level of e-government development than small, high income countries. A large country by land area, for example, must lay many more miles of fibre-optic cable than a small country to provide broadband connectivity to its citizens. Similarly, a country with a very large population must provide many more online access points – via kiosks, mobile phones, or other means – to its citizens than a country with a small population. Conversely, a country with a high income has more resources to apply to e-government development than a country with a low income.

For example, India has about 4000 times the population and about 130 times the area of Belize. Moreover, it has only about one quarter of the Gross National Income of Belize. As such, the effort required by India to provide e-government services is far greater than that of Belize.

The same is the case of China, which has about 15 times the population of Viet Nam, around 30 times the area, and some 30 per cent the income per capita. This implies that China has a lower availability of resources, skill levels and connectivity, at the margin, to devote to e-government development compared to Viet Nam, and that it must put forth a greater effort to achieve a similar level of e-government development. Despite these challenges the effort made by China has translated into a higher EGDI, at 0.5359.

The converse is true too. Countries endowed with a high income per capita, a small population, and high levels of connectivity face fewer challenges. It is to be expected that with fewer constraints, their e-government development efforts will be more developed so that e-services are available to all, and that high-income developed economies will go the extra mile to deploy ICT for transformation of societies for sustainable development.

This brings into sharp focus two aspects of e-government for development. Countries with a high per capita income, an established ICT infrastructure, and high levels of human capital can easily utilize these advantages to leverage the opportunity afforded by ICT and support sustained socio-economic development. However, in some cases, they may not be doing so fully. Others with lower levels shown by key indicators no doubt need a greater effort, but also have an opportunity to leapfrog long gestation developmental cycles by adroit utilization of ICT for development.

Regional Comparisons

Sustained integration, expansion and consolidation of government online offerings led to more than a 10 per cent increase in the world average of e-government development compared to two years ago. The region of Europe (0.7188) shows the highest e-government development followed by the Americas (0.5403).

Despite considerable strides towards bridging the digital divide, infrastructure and human capital limitations in several parts of the world impinge upon the ability of governments to spread – and the citizens to partake of – the benefits of information technology in the delivery of services. With a history of high levels of functional education and widespread telephony infrastructure, Europe and the Americas as a whole remain far ahead of the rest of the world regions. Asia, which is home to around three-fifths of the world citizens, has nevertheless only around 70 per cent of the level of e-government in Europe while the level of services in Africa barely squares off at 40 per cent of those in Europe. Within any region, countries at the lower percentile of e-development do not fare well either. This is especially true of the lower income countries in both Asia and Africa. The 10 least e-ready countries in Asia have barely 37 per cent of the level of e-government in Europe while in Africa the figure is little more than 20 per cent.

What is encouraging is the worldwide trend during the last decade. Since 2003 all regions of the world have steadily improved their e-government development offerings with European countries visibly taking of in the last two years. While some countries have advanced considerably over others, Asia as a whole progressed at a leaner rate till 2010 – almost in line with the advances in the world average – and then took of. With an almost flat curve for the period 2003-2012, e-government offerings in Africa advanced minimally, with the region as a whole still remaining least e-ready.

E-GOVERNMENT IN AFRICA

The key challenge for the e-government development of Africa remains the widespread lack of infrastructure and functional literacy. Despite recent expansion in mobile telephony, most countries in Africa remain at the tail end of the digital divide. These challenges have translated into a lower than world average e-government development for all sub-regions. Southern Africa (0.3934) consistently outpaces all other sub-regions. Though there has been some improvement in all sub-regions, except for Northern Africa and Middle Africa, it has been minimal, with the least e-ready sub-region being Western Africa (0.2171).

Africa has seen improvement in e-government with countries in the region looking to increase their online presence through developing websites for government ministries and agencies. Seychelles (0.5192) climbed several points to number one in the region in 2012 followed by Mauritius (0.5066) and South Africa (0.4869). It is notable that all of the African leaders increased their e-government development index value in 2012 but lost in comparative performance around the world, except for Kenya and Morocco, which gained in the world rankings from 124 to 119 and from 126 to 120 respectively. Tunisia (0.4833) and Egypt (0.4611) declined in rank substantially as did Cape Verde (0.4297) because their improvements did not keep pace with those of other countries around the world.

In 2012, Seychelles undertook further consolidation of infrastructure and its e-government development. Major improvements in mobile telecommunication and integration of thematic services in education, health, and finance with the national portal allowed it to improve its world ranking. Mauritius improved its offerings around 10 per cent with the national portal providing facilities for services such as appointments for vehicle inspections, scholarships and work permits. Although it ranked 2nd in the Eastern Africa region, its efforts could not keep pace in comparison to peers, leading to a decline in its global ranking.

Online services of Mozambique have gained ground in attempting to consolidate all information into one complete site, though this site lacks transactional services. Integrated services across sectors, including important legislation, are available. They include obtaining an identity card, registration of motor vehicles, finding private employment recruitment agencies, and the payment of taxes, to name a few. Progress on back-office integration can be found from the linkages to the various ministries and institutions of the government. Lack of infrastructure, especially broadband, remains a critical factor impeding the e-government efforts in other countries of this sub-region such as Mozambique and Rwanda as well, despite their progress in expanding services.

The top five countries in the Middle Africa sub-region all improved their service offerings in 2011. However the marginal improvement did not translate into rank improvements on a world level, except in the case of Cameroon (0.3070). The countries of Middle Africa trailed behind other countries of the world.

Gabon (0.3687) was the sub-regional leader followed by Sao Tome and Principe (0.3327) and then Angola (0.3203). Improved features such as Twitter and Facebook on the Gabon national site indicated a move towards greater participation and inclusion of the citizen. The national website of Sao Tome and Principe, though providing mostly static information, has archived data including sectoral information on health, education and the economy.

For effective e-government to materialize, planning and organization needs to accompany resource availability and an adequate level of human and physical infrastructure on the ground.

Though most countries of Northern Africa increased their e-government offerings since the last Survey, they slipped in overall world rankings this year primarily because other countries overtook them in infrastructural development, especially in mobile telephone access. Tunisia (0.4833) maintained its position as the leader of e-government in the sub-region. Morocco improved its e-government value (0.4209) reaching 120th. Algeria increased its e-government development value by 13 per cent and maintained its global rank. Egypt did not improve much and declined to 107th. South Sudan became the 193rd United Nations Member State and at the same time came online with a world ranking of 175th. At the same time, domestic political turmoil impacted upon the virtual presence of the Government in Libya (formerly the Libyan Arab Jamahiriya), which went off-line at the time of the survey assessment.

South Africa (0.4869) was the sub-regional leader, followed by Botswana (0.4186) and Namibia (0.3937). South Africa developed a solid presence covering many of the basic services and features while simultaneously developing transactional facilities and venturing into the networked presence stage. Though providing slightly higher online services than in 2010 and advances in mobile telephony, all countries in Southern Africa, except for Lesotho, fell behind due to continued low availability of infrastructure, especially for the use of broadband.

Cape Verde (0.4297) was the sub-regional leader. Although half of all countries, including Ghana (0.3159), Gambia (0.2688), Senegal (0.2673), and Liberia (0.2407) increased their offerings in 2012 all countries of the sub-region remained below the world average. Despite the upsurge in mobile telephony online in recent years, services in Africa remain circumscribed by lack of infrastructure.

E-GOVERNMENT IN THE AMERICAS

As part of their effort to advance citizen services, developed countries are paying greater attention to the concepts of an integrated government portal and the re-engineering of back-office processes in designing their e-government capabilities. E-government strategies are geared towards user-centric solutions, which serve to synergize governance processes and systems across multiple public administration domains.

The sub-region of Northern America (0.8559), encompassing only the United States and Canada, is the world leader with values far higher than the world average and all other sub-regions. In 2012, all sub-regions collectively improved performance in the Americas, including the Caribbean (0.5133) and South America (0.5507).

The top ranked countries in the Americas remained the United States followed by Canada, both of which were also among the world leaders. All countries of the region improved their e-gov-ernment in the past two years, which contributed to around 12 per cent improvement in the sub- regional average. The majority of the countries were also among the top 60 in world rankings.

The United States was found, as before, a best practice example of an integrated portal that provides easy to navigate design and collects and consolidates all information and services for citizens in one place, including agency services at the state and local level, which vastly increases the effectiveness of user search and uptake.

Barbados (0.6566) has been and remains the sub-regional leader among the Caribbean countries in 2012 followed by Antigua and Barbuda (0.6345) and the Bahamas (0.5793). The national site of Barbados offered a user friendly approach of "channels" such as the Government Channel, Citizens & was the leader with e-government offerings around 27 per cent higher than other countries of the sub-region. Closely following Mexico as number two in the sub-region, Panama (0.5733) improved its world ranking from 79 in 2010 to 66 in 2012. It is followed by El Salvador (0.5513) and Costa Rica (0.5397). On the other hand, even as mobile telephony increased in El Salvador, broadband and other access infrastructure remained low, impeding its online service delivery uptake. Other countries of the sub-region that improved e-services are also demonstrating that the expansion of mobile infrastructure has allowed them to complement, and indeed supplement, traditional access to narrow the digital divide.

Mexico upgraded its offerings in 2011 to include a comprehensive search service, which indexes federal, state and municipal web portals daily. With more than 400 million registries in its index, the national portal greatly expanded online services to citizens, including an open government initiative, special offerings for vulnerable groups, and a facility for the anonymous reporting of issues of concern to authorities. It allows for greater inclusion of the citizen through social media such as Twitter and Facebook and is among the select 19 per cent of world countries providing a single sign-on service.

Though Panama improved its online services, the main contributor to its advancement in this year's rankings is the expansion of mobile infrastructure, which is becoming an affordable technology among other countries as well. Panama has one of the highest penetrations of mobile subscribers in the region. As part of its Modernization Plan, to be completed in 2014, Panama is aiming to provide free access to the Internet for all citizens.

The United States (0.8687) leads this sub-region followed closely by Canada (0.8430). Since the United Nations Survey started tracking e-government development in 2003 both countries have been among the top world leaders with integrated closest to a pure integrated portal with access to interlinked searchable information from the United States Government, state governments, and local governments all in one place. Substantial back-office integration has gone into the user interface, which offers a simple convenient and easy-to-use facility for everything from government departments and agencies to verifying a social security number, getting an employer identification number, multiple online participation efforts and much more. Early recognition of the use of ICT for rolling out citizen portals and increasingly inclusive citizen services spread across theme, functionally and now by life cycle and events. For example, the United States e-government portal (ht p://www.usa.gov) comes centric services has contributed to the United States' top rankings in the last decade. Both the United States and Canada have consistently had e-government development levels far above the world average from 2003 to 2012.

Chile (0.6769) is the sub-regional leader in South America, followed by Colombia (0.6572). Whereas collectively the sub-region improved its e-government development by 13 per cent, of the 12 countries that make up this sub-region all declined in the world rankings except Brazil (0.6167) and Suriname (0.4344), indicating that countries within the region as well as around the world are investing in – and expanding – services faster than the countries of this sub-region.

E-GOVERNMENT IN ASIA

Asia is home to 60 per cent of humanity. With some Asian countries, including China and India, averaging around 8 to 9 per cent of the continent's GDP, Asia as a whole continued to expand e-government services further. Investments were made horizontally to expand infrastructure, including support for broadband and mobile access, while at the same time governments reached out to provide greater online services and improve e-governance. In 2012, three of the world's top 20 e-leaders are from Asia, and the region as a whole has a higher level of e-government development than the world average. While there has been improvement in providing e-services across the continent, some of the largest gains are found in Western Asia.

The Republic of Korea (0.9283), the world leader in e-government, is also the top performer in Asia with around double the average world e-government offerings. The 2nd slot is taken this year by Singapore (0.8474) followed by Israel (0.8100) and then Japan (0.8019). The performance of the

United Arab Emirates (0.7344) is especially notable as it advanced 21 positions to the ranking this year of 28th globally and 5th in Asia. The rapid progress of the United Arab Emirates is a best practice case highlighting how effective e-government can help support development. With double the population and three quarters of the GDP per capita, the United Arab Emirates has achieved around the same level of online services as those offered in Norway, a global leader at the 8th position.

Commensurate with global progress, all countries of Central Asia improved their service offerings, pulling up the sub-regional average by around 17 per cent. Kazakhstan was the sub-regional leader, improving its global ranking by around eight positions in 2012.

Kazakhstan in recent years has made efforts to modernize the public sector, including technology-based reform of administrative governance systems. A parallel effort has been a focus on the use of ICT for provision of services and inclusion. As in other developing countries the acceleration of informatization is aimed at increasing the efficiency of the government and exploiting synergies towards a sustainable model of development.

Ranked 2nd in the e-government development index in Central Asia, Uzbekistan has taken slow but significant steps toward increasing its online presence with the Government Portal of the Republic of Uzbekistan (ht p:/ /www.gov.uz). Though the country slipped in overall rank, it improved its services by around 13 per cent. If several years ago the website was merely informative and not at all interactive, government initiative and consequent legislative changes in 2007 and 2009 made possible a more comprehensive e-government portal. Compared to its earlier versions, the national site has added the following notable characteristics and features: effective organization of information; integrated archived information (laws, policies, etc.); an increased number of ministries linked to the portal; technical and web design features (RSS, audio, video, language, etc.) and static online downloadable forms.

The Republic of Korea offers around 87 per cent of all services assessed in the United Nations E-Government Survey 2012. The country's emergence as the world leader in information and communication technologies in fields such as broadband, semiconductors and third generation mobile devices – has guaranteed its fast growth and development in the area of e-government. Japan follows as number two in the sub-region, at 18th in world e-government development rankings. It is notable that all countries of Eastern Asia improved their service offerings and that the Democratic People's Republic of Korea came online.

Among others, China has made steady progress in overall e-government development. This is no small feat since it is a country of 1.2 billion people and a large land mass – both of which require more effort from the government, especially if the population is widely dispersed, than would a country with a small population living within a limited area. China has enhanced the quality of its government portal by providing comprehensive information, more integrated services across different sectors, and greater interactions between government officials and citizens.

All countries of Southern Asia fall in the lower half of the e-ready countries with approximately an equal number of them above and below the regional average. A low GDP per capita, a still evolving infrastructure and lower levels of functional literacy translate into low service provision and user uptake for the majority of the populations of India, Bangladesh, Bhutan, Pakistan and Nepal, with e-government development levels ranging from 0.2664 to 0.3829.

Maldives (0.4994) leads in the sub-region followed by the Islamic Republic of Iran (0.4876) and then Sri Lanka (0.4357). Service provision in Maldives builds on providing easy access to information to citizens and businesses. Through an 'I Want To' section organized by theme and life cycle, users can find information on service procedures, including how to obtain driver licenses, obtain ID cards and register vehicles. The national site of the Islamic Republic of Iran is available in two languages: Persian and English. Transactional service offerings in the Islamic Republic of Iran are joint public-private partnerships with some available online but also though banks and other local and national institutions.

With 1.2 billion people and challenges associated with a large population, e-services in India are in the formative stage. The Government of India has made substantial efforts in the last few years to overcome the challenges, including that of connectivity to its 70 per cent rural population. Looking towards sustainable growth the government has announced that Rural Broadband Connectivity to all 250,000 Panchayats (local governments) in the country will be provided in three years to bridge the digital divide.

Other countries in the sub-region such as Pakistan are also providing more e-services. In line with the Government of Pakistan's policy to digitize e-services, the Multi-biometric e-Passport project aims at improving transparency in the public sphere.

Tenth in the world ranking, Singapore (0.8474) is the leader in the South Eastern Asia region and a best practice example. It is among the vanguard countries employing advanced technology for innovative future solutions, organizing information in a seamless structured and user-targeted manner. Malaysia (0.6703) continues to be the 2nd leading country in the South Eastern Asia region in 2012, not least because of its impressive service provision through an A-Z topics section as well as a life cycle feature targeting the user as a child, teenager, adult or elderly person. An impressive national health portal, *MyHealth*, uses ICT to inform citizens on health issues in Malaysia.

Like other leading countries, the key to the advancement of Israel, the leader in Western Asia, is its integrated approach to e-government development. Israel has added 10 points to its ranking, advancing from 26th position in 2010 to 16th in 2012.

Close behind are Saudi Arabia (0.6658) and Qatar (0.6405), both of which have undertaken to expand citizen centric services as reflected in their notable performance, which raised their global rankings to 41st and 48th respectively. The principle goals of the Saudi Arabian e-government offerings are to raise the productivity and efficiency of the public sector, increase the return on investment in ICT and provide easy-to-use, timely accurate services. A separate e-payment portal has been developed through which citizens are able to handle all online transactions.

In accordance with the strategy of the Supreme Council of Information and Communication Technology (ictQATAR), Qatar developed a strategic plan for the implementation of an integrated government programme. The plan aims to provide an enabling ICT environment through legislation, policies, guidelines and standards, such as the e-commerce policy, security policy, and data protection policy. It addresses ICT 'readiness' by offering e-services through a unified system of government networks; provides a centralized, secure, government data centre and a payment platform; and seeks to improve 'usage' by increasing the number of e-services such as businesses' registration, visa services and resident permits. Hukoomi, the national portal of Qatar, integrates back-office processes to allow easy access to over 100 topics and articles with detailed information about Qatari law and society. Online services available include, among others, payment for utilities, renewal of health cards, settlement of traffic violations, visa applications, and licenses.

E-GOVERNMENT IN EUROPE

Europe as a region has been in the vanguard of information technology and setting the pace for others to follow. Building on the existing strength of high levels of human capital and infrastructure, the transformative role of ICT has been recognized and adopted to further streamline e-government services. Moving beyond improving public sector efficiency, Europe is now looking to adapt innovative technologies to human development and economic sustain-ability in the future.

The European region has the highest level of e-government development, which is around 50 per cent higher than that of the world as a whole. Western and Northern Europe offer the most online services but considerable gains were made by Southern and Eastern Europe as well in 2012.

With a common e-government framework, EU countries are encouraged to deploy advanced technologies, institute bet er governance and provide expanded services with concomitant pursuit of greater transparency, efficiency and inclusion. Notwithstanding, differences remain between regions and within them. Key European countries spend more than double the EU average amount per capita on ICT; others, around half of it.

The Netherlands (0.9125) made substantial gains, advancing to the top position in Europe and 2nd in world rankings, followed by the United Kingdom (0.8960) in 3rd place and Denmark (0.8889), which also advanced and occupies the 4th position this year. Within the aforementioned common e-government framework, all of the top countries of Europe offered more or less the same level of user centric services to their citizens resulting in marginal assessment difference among them. For example, Germany (0.8079), the 10th leading country in Europe as a whole, achieved about 89 per cent of the e-government development level of the regional leader, the Netherlands.

Advancing 32 positions in the world rankings, the Russian Federation (0.7345) became the leader in Eastern Europe followed by Hungary (0.7201) and the Czech Republic (0.6491).

As in other parts of Europe, all countries of the sub-region improved their e-government development in 2012, advancing the sub-regional average by 16 per cent even though they could not maintain their rankings, except for Belarus and the Russian Federation. Being the largest country in the world and consisting of eight federal districts influences the development of e-government in the Russian Federation. The Government recently announced investments of around 80 billion rubles for the informatization of federal government bodies and other initiatives related to the development of e-government. In Hungary, the focus of programmes to develop the information society encompassed support for improving ICT skills in the labour market, targeting small and medium enterprises and increasing the number of ICT experts.

E-services were increasingly the norm in other countries as well. In 2010, Moldova, another country that improved its ranking, in collaboration with the World Bank, started implementation of a *Strategic Programme for Technological Modernization of the Government*, aimed at ICT-led institutional reform to increase access to information and promote digital services (e-services), and to improve public administration. By improving the governance frameworks the programme is expected to result in simplification and consolidation of administrative procedures leading to transparency and effectiveness.

Six out of ten countries of Northern Europe were among the world leaders. Number three in the world, United Kingdom (0.8960) was the leader in Northern Europe followed by Denmark (0.8889 – 4th) and Sweden (0.8599 – 7th).The focus on integration of back-office departments for improvement of user interface is evident in the integrated services on the United Kingdom's *DirectGov* (ht p://www.direct.gov.uk), which provides a one-stop-shop for all government information and services. Its comprehensive 'Do It Online' page lists all public services, forms, tools and transactions that the government provides in a user-friendly manner. The *Jobcentre Plus* page located on the portal allows citizens to search one of the United Kingdom's largest online databases of job vacancies.

With a focus on service delivery, government national portals are organized according to domains in most of the EU countries. Denmark's overall strategy appears to go well beyond simply providing a single portal. Rather, it seems to be focusing on multiple entry points to government based on various interest groups and constituencies. Early adoption of online transactional services has resulted in substantial cost efficiencies. In Denmark, for example, electronic invoicing saves taxpayers •150 million and businesses •50 million a year. According to one estimate, similar efforts all across the EU would result in annual savings of around •50 billion.

In Northern Europe, the progress of Finland was especially noteworthy as it gained 10 positions to rank 9th in the world. The National Knowledge Society Strategy 2007-2015 in Finland focuses on the provision of multichannel, interactive e-services together with interoperability of information systems in the public administration. The *suomi.fi* (ht p:// www.suomi.f /suomif /suomi) portal provides a single access point to online public services offered from both state and local authorities, organized around daily life events.

Some other Northern European countries also fortified their e-services, providing greater access and inclusion to citizens. Though they did not maintain their global ranking, other countries such as Lithuania (0.7333 – 29th), Ireland (0.7149 – 34th), and Latvia (0.6604 – 42nd) also improved e-government applications, networking, and other web services. The e-government gateway of Lithuania (ht p://www.ep-aslaugos.lt) has been developed under the auspices of the Information Society Development Commitee for the purpose of providing seamless public administration services and information to residents and business. Covering content in both the Lithuanian and English languages, the gateway provides 211 first-level public services, 167 second-level services, 33 third-level services and 12 fourth-level services. The one-stop-shop portal of Ireland approaches the delivery of e-services in such a way that enables users to tailor those services to their needs at a place and time that suits them, together with an overview of the extent of public services online. Latvia's one-stop-shop portal (ht ps:/ /www.latvija.lv) offers 29 e-services and online banking (e-payments). Points of Single Contact allow service providers to obtain information through a single entry point and complete required administrative procedures electronically in order to commence provision of services in a chosen business sector in Latvia. An advanced users authorization system through bank or ID cards ensures users privacy and security. The online banking system allows users to employ the system more efficiently and securely (e.g., for income tax declaration).

Even though its global ranking dropped, Spain (0.7770) remained the leader in Southern Europe, followed by Slovenia at 25th and Croatia at 30th in world rankings. The national site of Spain is available for the user in five languages with information services and easy-to-navigate features. In Slovenia the public sector reforms have included digitization of governance processes and services for improved functioning. The State portal of the Government of Slovenia (ht p://e-uprava.gov.si) organizes information to citizens by links to 18 life event categories such as work and employment, health and social affairs, personal finance and taxes, environment, education and youth, and social welfare so that information on government services can be easily and quickly accessible.

Among other countries that increased their world rankings are Italy, Portugal, Greece and Serbia. Particularly notable is Serbia's performance, which advanced 30 positions to arrive at 51st in the world rankings. In Serbia, the Digital Agenda Authority is responsible for introducing online services to improve economic efficiency and citizens' quality of life, and for implementing e-government in accordance with a "one-stop-shop" principle. Among other initiatives, the Authority created Serbia's e-services portal, eU-prava (http://www.euprava.gov.rs), which aggregates services and information from more than 27 governmental authorities, including municipal authorities.

Of the countries of the region which are global leaders, several offered examples of best practice. In the Netherlands, efficiency and citizen inclusion are the objectives of the e-government strategy. Integration of a back-office management system has been undertaken with a belief that citizens should provide information once. The government is building an e-government infrastructure encompassing citizen access to government processes including electronic authentication, uniform identification numbers for both citizens and businesses and electronic personal identification. As part of its broader ICT strategy the focus of e-government in the Netherlands was on improving efficiency of services concomitant with reduction of administrative cost and burden.

Based on extensive technological infrastructure, the recently concluded National Implementation Programme (NUP) for Better Services and e-Government laid out agreements among the national government, provinces, and municipalities to improve service delivery. Its high levels of broadband connectivity ensured further enhancements in e-services undertaken during the last few years.

Australia continues to be the leader in the Oceania region. The national portal (http://aus-tralia.gov.au) acts as a one-stop-shop that connects citizens to the information and services of around 900 government websites and state and territory resources. Information can be quickly and easily accessed through the 'People' and 'Topics' sections, which categorically filter specific content while the 'Services' section allows citizens to perform many functions such as making payments for taxes, driver license renewals, vehicle and business registrations, lodging online forms and making online inquiries. The integrated portal of the Government of New Zealand provides a one-stop-shop portal for information, images and resources from all New Zealand government agencies and government funded sites.

E-Government in Least Developed Countries

The least developed countries (LDCs) group was led by Samoa (0.4358) at 114th place, followed by Tuvalu (0.3539) which, at 134th, made considerable progress since 2010. Notable advances were also made by Vanuatu (0.3512) and Rwanda (0.3291), which moved up to 135th and 14 0 th, respectively. Overall, the LDCs remain hampered by a lack of infrastructure, both physical and human. Despite advances in mobile communication lack of functional skills limit user uptake.

Though there is considerable progress in the expansion of online services, one of the primary challenges that remains is integration of back-end processes with efficient, user friendly, and target-oriented services delivery. Countries around the world are increasingly adopting integrated, multichannel and user-centric services online. Though efforts towards deploying ICT for sustainable development are evident, the extent, design and approach to user interface vary depending on several factors, including leadership and vision, planning and organization, level of income and absorptive capacity in the country.

References

Ashok Agarwal and V Ventaka Ramana. *Foundations of e-Government*. ICEG'07 5th International Conference on e-Governance

Cordella, A (2007), E-government: towards the e-bureaucratic form?, *Journal of Information Technology*, 22, 265–274.

Jane Fountain, (2001). *Building the Virtual State: Information Technology and Institutional Change*

Kushchu I. and Kuscu M. H. (2003). "From e-Government to m-Government: Facing the Inevitable". *The 3rd European Conference on e-Government*. pp. 253–260.

West, Darrell. (2008). *State and Federal Electronic Government in the United States*. The Brookings Institution. Retrieved on 2008-09-16.

6

E-Democracy and Citizen Participation

Webster's defines democracy as "a government in which the supreme power is vested in the people and exercised by them directly or indirectly through a system of representation." Putting an "e" in front of democracy means nothing more than using information technology tools to facilitate, improve and ultimately extend the exercise of democracy.

E-democracy has both a tactical side and a strategic side. On the tactical side, information technology has advanced communication and the access to information arguably better than any known medium. But something even more fundamental is at hand. The underlying core principle of democracy is an informed and engaged citizenry. Most governments get passing marks for "informing" citizens via digital communication. But the vast majority have a long way to go to "actively engage" citizens or to effectively exert global influence using digital media. These elements comprise the most overlooked dimension of e-democracy – the strategic side. How can a government use digital media to both actively engage citizens and advance its public policies to the world community?

Engaging your "own" citizens or constituents through digital media includes enhancing active participation in law-making, policy-making, and legislative process, all of which are influenced by a variety of forces— public opinion, debate, lobbyists, special interest groups, consultation with constituents, committee hearings, and expert testimony. The regulatory process (subsequent to enactment of law) follows many of the same communicative and collaborative patterns as law making. Lest we not forget, the ability to leverage digital technology by political parties, campaigns and candidates is also part of the equation. Voter registration, election or referenda voting, and on-going communication between constituents and their elected representatives are equally integral to e-democracy.

Despite an increasingly digital world, attention must be directed toward the digital divide. A 2002 national survey in India revealed that fewer than one percent of adults had used the Internet in the preceding three months. In response, a growing national network of owner-operated computer centers has emerged that may eventually serve up to 700 million people in 600,000 rural villages. Village entrepreneurs who provide cybercafé services tap into wireless technology through India's fiber-optic network which reaches 85 percent of the country, compensating for the lack of access to telephone land lines. For small fees, citizens can access government officials, records and online medical consultations. Other services already include education, commerce, and participatory democracy.

The Internet has greatly lowered the costs of transmitting information, enabling people to bypass traditional intermediaries whose power revolved around the control of information: national governments, the diplomatic corps and transnational corporations, among others. Non-governmental organizations (NGOs), ethnic communities, individuals, and, yes, even terrorists, use the Internet to create global platforms and political influence.

There is a huge difference between information/communication within one's own borders and winning the hearts and minds of those in other cultures. Affecting world opinion through information technology may not be as obvious, but has far-reaching implications. The Aspen Institute defines "Netpolitik" as the exploitation of powerful Internet capabilities to shape politics, culture, values, personal identity…and public perception." They further define "soft power" as how a government uses persuasion, public information, education, communications, culture, trade, aid, investment and marketing to secure public support of its interests, values and policies.

As parties to the democratic process, all government entities (international bodies, governments, political parties, elected representatives) need to urgently develop new and effective public and global communication skills using digital media.

As the former U.S. Secretary of State, Madeleine Albright and Robert Hormats, vice chairman of Goldman Sachs and a former top official at the

State Department and Office of the U.S. Trade Representative noted at the 2003 Aspen Institute Communications and Society Program Roundtable, it's not just the volume of information it's also the velocity of information. In the past, "...diplomatic communications were carried on through predictable venues and stable deliberative processes. The circle of knowledgeable participants was well established. The number of participants with access to accurate, timely information was relatively small." Not anymore. With CNN and the Internet, coupled with privatization of mass communications, the sheer volume and speed of information forces instant reaction and decision-making. Governments can no longer rely solely on formal intelligence reports, diplomatic cables and in-house experts. Survival dictates a strategy to effectively influence input into digital media channels and dynamically monitoring and interpreting digital output of other entities.

E-Democracy Models

Academics and others are beginning to explore models of e-democracy. The Organisation for Economic Co-Operation and Development (OECD) defines three types of e-democracy interaction – one-way information provision; a two-way relationship where citizens have opportunity to give feedback on issues; and, finally, a partnership relationship whereby citizens are actively engaged in policymaking. Similarly, Coleman and Gotze suggest four scenarios. The first is technology supporting direct democracy. The second encompasses online grass-roots civic communities of interest. The third addresses online surveys and opinion polls. The fourth points to technology as a way to engage citizens in policy deliberation. Gartner Dataquest has developed a four-stage model of "e-development" which applies to citizen services as well as e-democracy. In the first stage (presence) the Internet site provides information online in a static format. In the second stage (interaction) citizens search information, download forms, or access links to other relevant sites. The third stage (transaction) alleviates the need to complete a transaction by mail or make an office visit. The fourth stage is transformational. Some attributes of this stage are wireless access, enabling sites to push government information to citizens and robust customer relationship management (CRM) tools. Inherent in this stage is a redesign of workflow and processes.

Consensus is clearly beginning to emerge, at least regarding e-democracy as it affects citizen engagement within borders. Progression is toward facilitating proactive citizen engagement that can influence and improve policy making—not simply another form of citizen communication. However, these models fall far short of a strategic vision of e-democracy.

The Institute for Electronic Government's model takes a leap forward in both the definition and implementation of e-democracy (Figure 1). The IEG model is not limited to the citizen-to-government point of view, mapping progression from an informed to an engaged citizenry. It also serves as a scorecard of digital savvy—how successfully a government entity interprets and responds to the digital world and exploits technology accordingly to advance influence. That 'entity' might be an elected representative, a legislative body, a provincial or national government, a political party, or international organization. The model helps leaders think through how to fold both tactical and strategic e-democracy efforts into an overall e-government strategy. With one glance, a government can identify its current position against characteristics at various sophistication levels and see what e-initiatives can take them to the next level. One axis measures the degree of engagement. The other axis measures influence.

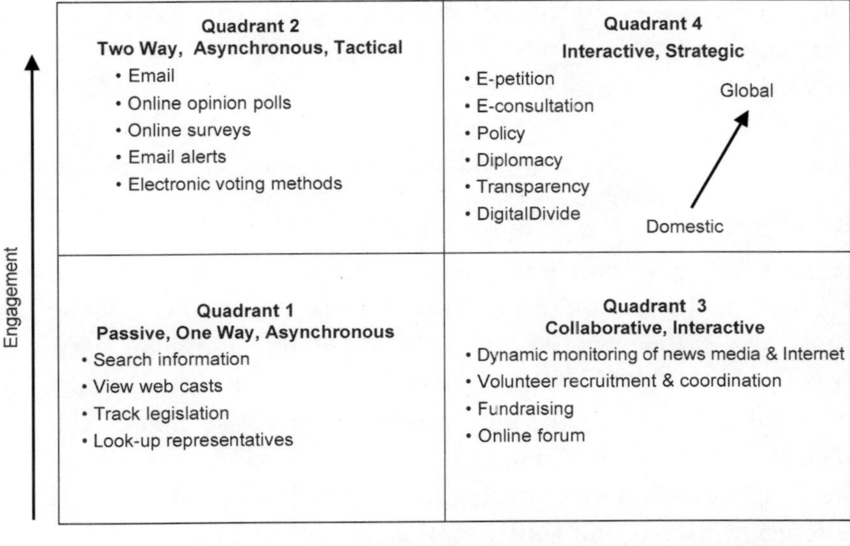

Figure 1. E-Democracy Model Institute for Electronic Government

Quadrant 1.

Most government entities—governments, legislative bodies, international organizations, political parties—have done a pretty good job of making information available online. That's a fundamental step in e-democracy tactics. According to the recent Pew Foundation's Internet & American Life Project:

- 68 million Americans visited a website, up from 40 million in just twelve months.
- 42 million Americans used government web sites to research public policy issues
- 23 million Americans have used the Internet to send comments to public officials about policy choices.
- 13 million Americans participate in online lobbying
- 14 million have used government Web sites to gather information to help them decide how to cast their votes. 13 million have participated in online lobbying campaigns.

Legislatures have begun to understand not only how to use technology to communicate with constituents but how to operate as modern businesses exploiting technology. Evidence of the growing awareness and importance of technology, the Council of State Governments reports that over the past four years thirty four U.S. states have provided laptops or PCs in legislative chambers linking representatives to party leadership, to legislative systems, and to their constituencies. Many regularly web cast and archive proceedings.

Quadrant Two

Entities in this quadrant have made great strides to open two-way communication. Let's look at this from the citizens' point of view. Who exactly is "*the government*" anyway? As they say in the film, Ghostbusters, "Who you gonna call?" Is it the city councilman or mayor? The appointed school board member? Maybe the governor. Could it be a state legislator, Member of Parliament, the Chair of a Senate Sub Committee? Is it a county or federal agency? Is it 10 Downing, the White House? Increasingly, is it the European Union or the United Nations?

The correct answer, of course, is "all of the above." We are citizens of towns, cities, states, provinces, countries and, yes, the world. That's the challenge. Every public institution and those who serve in them are obliged to move beyond information dissemination to open two-way communication channels relevant to the digital age in which we live.

Although entities in Quadrant Two may have achieved two-way capability, its nature is still largely asynchronous. The Congressional Management Foundation reports that in 2001, United States Senate offices received as many as 55,000 emails a month. Some congressional offices report that up to 60% of all correspondence is through email. Even so, the report notes 75% of House members respond to constituent email with postal letters!

The M.I.T. Artificial Intelligence Laboratory has explored managing public access and participation via the Internet in government inquiry and regulatory processes. Because the Internet broadens and cheapens access to these processes, it can dramatically increase the number of responses to proposals. Fighting fire with fire, overload can be addressed using a variety of technologies to manage technology – such as software to sort and respond to email.

Inviting citizens to sign up for email alerts on various issues is one proactive communication strategy & reduces the volume of unnecessary individual email. Many times a citizen just wants to tell government what they think, not necessarily expecting a response. For example, U.S. Senator Don Nickles (R-OK) reduces unnecessary e-mail volume on hot topics through a pre-emptive strategy. "You Called It," is an innovative feature on his Web site that lists the top five issues constituents contacted him on in the past week and his position on each. This way, constituents who want to know the Senator's view don't have to contact the office— it's easily found on his home page.

As one of its Homeland Security initiatives, Arlington County, Virginia launched a citizen emergency alert system (www.arlingtonalert.com) to contact residents in the event of an emergency. The system sends alerts, updates, and notifications to ALL of a citizen's listed devices and email accounts, including cell phones, pagers, Blackberries, or other PDAs. Messages are sent in English or Spanish. During the regional Hurricane Isabel emergency and resulting widespread power outages, use of this communication tool skyrocketed.

The City of Fairfax, Virginia, has an extensive e-message alert system for both emergency and non-emergency communication. The City pushes information to registered citizens (by either email or phone) including reminders of property tax due dates, road closures, school closings, city events, and weather alerts.

The town of Issy-les-Moulineaux, near Paris, regularly polls a representative sample of 650 residents online on a variety of local issues – public safety, schools, and urban development. This consultation must now precede any major local project and gives the Town Council a tool to help in decision-making.

Quadrant 3

This quadrant extends interactive capability. Although still largely asynchronous, communication begins to evolve into collaboration. Most visible in this stage are political players and the electoral process with tactics such as recruiting and organizing volunteers online, online fundraising, campaigning, communication with constituents and the media, voter registration and voting.

In the United States, over three thousand counties currently deploy voting at over 200,000 polling sites. Technologies range from punch cards, optical scanning, lever, direct recording electronics, to paper ballots. Many of these are aging or obsolete systems. Few, if any, standards exist, even within states. Numerous countries have employed various forms of new electronic voting methods at both the polling place and outside the polling place, most notably Brazil, Australia, the UK, and Switzerland. These initiatives include touchscreen voting machines, interactive voice response (IVR) technology, PC-based systems, public kiosks, interactive digital TV, and voting using handheld mobile devices via short message service. However, these electronic voting devices have had their share of controversy, including questionable auditability, accountability, reliability, recount capability, and lack of permanent records or uniform standards. Very few governments have ventured past the pilot stage of voting over the Internet. In January, 2003, the small Swiss village of Anieres outside Geneva held its first legally binding Internet vote.

The collection and counting of votes is only one part of the challenge. Many times, changes made to traditional voter registration systems (such as address changes) are not processed in time for election day. Redundant voter data may exist in several locations within a state (if voter moves). These are straightforward technology issues—database design and integration—relatively easy and inexpensive to correct.

Quadrant 4

Domestic Citizen Engagement

Managing the policy making process is not unlike managing product life cycle. The earlier in the policy making cycle, the more likely citizens can influence the outcome. The Hansard Society in the UK contends that mechanisms for promoting public deliberation and embedding it within the constitutional process and demonstrating real links between public input and policy outcomes need to be devised. They highlight five reasons for governments to do so: (1) improve the quality of policy by tapping wider sources of expertise under conditions of increasing complexity; (2) prepare for greater and faster interactions demanded by information society; (3) integrate public input into policy making; (4) respond to calls for transparency and accountability; (5) strengthen trust in government.

Quadrant Four represents the highest level of e-democracy sophistication at least for the foreseeable future – strategic, interactive, synchronous, and global in nature. Democratic institutions should at least actively pilot initiatives in these areas now. If there is any doubt that leading-edge governments are already exploiting technology to gather input from citizens and businesses to subsequently determine a course of action accordingly, one only needs to visit a few sites to identify this trend.

On the Queensland, Australia website (www.qld.gov.au) citizens have an impressive array of opportunities to interact with the government. The "Get Involved With Government" choice links citizens to their representatives, to Queensland agencies, and to Parliament.

The 'Queensland Agencies' link gives citizens background information on an issue, current law or proposed legislation and invites direct citizen comments which goes to committee and then eventually to Parliament to help formulate policies and standards on a variety of legislative issues. The 'Queensland Parliament' link empowers a citizen to make a formal, direct request to Parliament in the form of an e-petition with the object of "persuading Parliament to take some particular action." Citizens can also review existing e-petitions and add their own signatures in a show of support, or express their objections. The site also surveys users about the e-petition process itself-a built-in quality control & improvement mechanism. Likewise, the Scottish Parliament was an early innovator in e-petitioning.

Citizens can create an e-petition or comment or add their support to an existing e-petition-all electronically (www.scottish.parliament.uk/e-petitions/index.htm). The International Teledemocracy Centre, founded in 1999 by Scotland's Napier University and BT Scotland aims to develop and apply advanced information and communication technology to enhance and support the democratic decision-making process. Their mission includes:

- Promoting the application of Information and Communications Technologies (ICT) by governments and parliaments worldwide in order that elected members and supporting staff can conduct their
- business more effectively and efficiently.
- Demonstrating how technology can contribute to more openness and accessibility in government.
- Encouraging and assisting the public, voluntary organizations and business to participate in government through the use of technology.

In 2001 the European Commission adopted an "Interactive Policy Making" (IPM) project to improve the European Union's governance. Through its website, "Your Voice in Europe" (http://europa.eu.int/ yourvoice), the IPM collects and analyzes citizen and business input to evaluate existing EU policies and to solicit consultations on new initiatives. The purpose is to make EU policy-making more transparent, comprehensive and effective, giving stakeholders an active role in the policy making process.

Global Positioning of Democracy

The focus of most e-democracy efforts to date have been largely directly internally to stakeholders, a jurisdiction or a domestic audience. That is clearly appropriate to communicate and engage citizens. On the other hand, the Internet knows no borders and harbingers of change are emerging.

In 2001, Japan's Prime Minister Koizumi launched the "Koizumi Cabinet Mail Magazine." This email magazine is distributed weekly from his office to explain his policy positions. More than two million subscribers now participate. For the anniversary celebration of the 100th edition, twenty subscribers were selected by lottery to meet in person with the prime minister to discuss a broad range of issues. Although only available in Japanese at this time, the email magazine initiative is an intriguing step toward a more strategic view of e-democracy that extends beyond Japan's borders.

Canada gets high marks for effectively advancing its public policies to a global audience (www.canada.gc.ca). This site probably represents early state-of-the-art in terms of explicitly expressing Canada's perspective on a variety of world issues directed to *non-Canadians*. Canada's website has a link for 'Non-Canadians' where people can further link to 'Canada and the World.' Here, content ranges from Canada's view on foreign policy, such as peacekeeping and counter terrorism, to the country's position on humanitarian aid, arms control, science and technology.

There are countless international websites for indigenous peoples, activist groups, grassroots organizations, and others who employ a variety of influence techniques ranging from information portals to activism. Government entities have a lot to learn from these early adopters to advance domestic and foreign affairs, enhance security, and promote democracy.

Over the next decade, e-democracy efforts will start to bear long-awaited fruit. But it won't happen without reasoned and deliberate action. First, leaders need to understand the importance e-democracy in their role as statesmen in a digital world. Secondly, leaders need to understand why their information technology infrastructures are essential to this vision. Government entities invest in information technology infrastructures for a variety of purposes. E-democracy should be one of those driving forces. Most leaders recognize technology infrastructure is the enabling foundation for internal government transformation and a vehicle to provide government services to citizens. The very same technologies can support sweeping changes in e-democracy. It's time to graduate from the now commonplace delivery of information and government services online to a more strategic view that promotes the philosophy and practice of democracy in the free world.

Strategies for Governance Processes

E-democracy is a form of direct democracy that uses information technologies and communication technologies and strategies for political and governance processes. Such processes can be used for governance of local communities, nations and internationally. Democratic actors and sectors in this context include governments, elected officials, the media, political organizations, and citizens/voters.

E-democracy aims for broader and more active citizen participation enabled by the Internet, mobile communications, and other technologies in today's representative democracy, as well as through more participatory or direct forms of citizen involvement in addressing public challenges.

E-democracy is a relatively new concept, which has surfaced out of the popularity of the Internet and the need to reinvigorate interest in the democratic process. Access is the key to creating interest in the democratic process. Citizens are more willing to use Web sites to support their candidates and their campaign drives. In the United States, just over 50% of the population votes, and in the United Kingdom, only 69% of citizens vote.

The research indicates that the political process has been alienated from ordinary people, where laws are made by representatives far removed from ordinary people. The goal of e-democracy is to reverse the cynicism citizens have about their government institutions. However, there are increasing doubts about the real impact of electronic and digital tools on citizen participation and democratic governance, and warning against the "rhetoric" of electronic democracy.

There has been a significant growth in the last four years, and implementation rates have topped out in many of the categories. To see this data, go to NIC Resources website. Public-and private-sector platforms provide an avenue to citizen engagement while offering access to transparent information citizens have come to expect.

To develop these public-sector portals or platforms, governments have the choice to internally develop and manage, outsource or sign a self-funding contract. The self-funding model creates portals that pay for themselves through convenience fees for certain e-government transactions. Early players in this space include govONE Solutions, First Data Government Solutions and NIC, a company built on the self-funded model.

Social networking is an emerging area for e-democracy. The social networking entry point is within the citizens' environment, and the engagement is on the citizens' terms. Proponents of e-government perceive government use of social networks as a medium to help government act more like the public it serves. Examples of state usage can be found at The Official Commonwealth of Virginia Homepage, where citizens can find Google tools and open social forums.

Government and its agents also have the opportunity to follow citizens to monitor satisfaction with services they receive. Through ListServs, RSS feeds, mobile messaging, micro-blogging services and blogs, government and its agencies can share information to citizens who share common interests and concerns. Government is also beginning to use Twitter. In the state of Rhode Island, for instance, Treasurer Frank T. Caprio is offering daily tweets of the state's cash flow. many state agencies also "tweet"

Practical Issues

A number of practical issues surround e-democracy. In the media, on the Internet, and in popular consciousness, there is a strong and generally unchallenged view that the Internet is the new electronic cradle of democracy. The original source of this view is probably the relatively unfettered speech found in Internet newsgroups, mailing lists, blogs, wikis and chat rooms.

The Internet currently does have several attributes that encourage thinking about it as a democratic medium. Part of this can be traced to the design principles that were established early in its evolution. The lack of centralized control suggests to many people that censorship or other attempts at control will be thwarted. Other attributes are a result of social design in the early days, the strongly libertarian support for free speech, the sharing culture that permeated nearly all aspects of Internet use, and the outright prohibition on commercial use by the National Science Foundation, for example.

The Internet's most significant contribution was the idea of unmediated many-to-many communication on a large scale, through newsgroups, chat rooms, MUDs, and many other modes. This type of communication ignored the boundaries established with broadcast media, such as newspapers or radio, and with one-to-one media, such as letters or landline telephones. Finally, because Internet is a massive digital network with open standards, universal and inexpensive access to a wide variety of communication media and models could actually be attained.

Some practical issues involving e-democracy include: effective participation; voting equality at decision stage; enlightened understanding; control of the agenda; and inclusiveness

Citizens' Roles

The Internet provides a distinctive structure of opportunities that has the potential to renew interest in civic engagement and participation. Civic engagement can be understood to include three distinct dimensions: political knowledge (what people learn about public affairs), political trust (the public's orientation of support for the political system), and political participation (conventional activities designed to influence government and the decision-making process).

The information capacity available on the Internet allows citizens to become more knowledgeable about government and political issues, and the interactivity of the medium allows for new forms of communication with government and elected officials. The posting of contact information, legislation, agendas, and policies makes government more transparent, potentially enabling more informed participation both online and offline.

Internet as a Campaign Tool

The Internet is viewed as a platform and delivery medium for tools that help to eliminate some of the distance constraints in representative democracy. Technical media for e-democracy can be expected to extend to mobile technologies, such as cellphones.

Most importantly, the Internet is a many-to-many communication medium, whereas radio and television, which broadcast few-to-many, and telephones, which broadcast few-to-few, are not. Also, the Internet has a much greater computational capacity, allowing strong encryption and database management, which is important in community information access and sharing, deliberative democracy and electoral fraud prevention. Further, people use the Internet to collaborate or meet in an asynchronous manner—that is, they do not have to be physically gathered at the same moment to get things accomplished.

Using the Internet as a political campaigning tool has become a cheaper and more convenient alternative for many politicians, in comparison to traditional door-to-door knocking or telephone campaigning. Candidates are also beginning to use social networking sites to reach younger audiences, creating potential supporters to campaigns. E-mail chains and political blogs also have had a major impact with online campaigning. Views are expressed by adding comments to political blogs or web pages. Point-and-click advertising (interactive advertising online) also has influenced traditional mail or television campaigning.

The lower cost of information exchange on the Internet, as well as the high level of reach that the content potentially has, makes the Internet an attractive medium for political information, particularly amongst social interest groups and parties with lower budgets.

For example, environmental or social issue groups may find the Internet an easier mechanism to increase awareness of their issues, as compared to traditional media outlets, such as television or newspapers, which require heavy financial investment. Due to all these factors, the Internet has the potential to take over certain traditional media of political communication, such as the telephone, the television, newspapers and the radio. The civil society has gradually moved into the online world.

Another example is openforum.com.au, an Australian non-profit eDemocracy project that invites politicians, senior public servants, academics, business people and other key stakeholders to engage in high-level policy debate.

Novel tools are being developed that are aimed at empowering bloggers, webmasters and owners of other social media, with the effect of moving from a *strictly informational* use of the Internet to using the Internet *as a means of social organization* not requiring top-down action. Action triggers, for instance, are a novel concept designed to allow webmasters to mobilize their viewers into action without the need for leadership. These tools are also utilized worldwide: for example, India is developing an effective blogosphere that allows Internet users to state their thoughts and opinions.

Electronic Support for Local Democratic Groups

Citizens' associations play an important role in the democratic process, providing a place for individuals to learn about public affairs and a source of power outside that of the state, according to theorists like Alexis de Tocqueville. Public-policy researcher Hans Klein at the Georgia Institute of Technology notes that participation in such forums has a number of barriers, such as the need to meet in one place at one time. In a study of a civic association in the northeastern United States, Klein found that electronic communications greatly enhanced the ability of the organization to fulfill its mission.

There are many forms of association in civic society. The term *interest group* conventionally refers to more formal organizations that either focus on particular social groups and economic sectors, such as trade unions and business and professional associations, or on more specific issues, such as abortion, gun control, or the environment. Other traditional interest groups have well-established organizational structures and formal membership rules, and their primary orientation is toward influencing government and the policy process. Transnational advocacy networks bring together loose coalitions of these organizations under common umbrella organizations that cross national borders.

The Internet may serve multiple functions for all of these organizations, including lobbying elected representatives, public officials, and policy elites; networking with related associations and organizations; mobilizing organizers, activists, and members using action alerts, newsletters, and emails; raising funds and recruiting supporters; and communicating their message to the public via the traditional news media.

Information and communications technologies are neither democratic nor undemocratic; they are merely means to an end and not normative by their nature. They are tools that may be deployed to achieve certain goals (e.g., contradictory goals, such as coercive control or participation can be fostered by digital technology). However, certain institutional framework conditions may either support or hamper the use of electronic means for the benefit of democratic processes. Risks and opportunities of the digitization of democratic processes depends therefore to a large extent upon the particular institutional framework conditions of the chosen democratic model (which is mainly set out in the Constitution, including the type of the underlying social contract, specific aspects of the rule of law, representative democracy or direct democracy, etc.).

Contemporary technologies, such as electronic mailing lists, peer-to-peer networks, collaborative software, wikis, Internet forums and blogs, are clues to and early potential solutions for some aspects of e-democracy.

A number of non-governmental sites have developed cross-jurisdiction, customer-focused applications that extract information from thousands of governmental organizations into a system that brings consistency to data across many dissimilar providers. It is convenient and cost-effective for businesses, and the public benefits by getting easy access to the most current information available without having to expend tax dollars to get it. One example of this is transparent.gov, a free resource for citizens to quickly identify the various open government initiatives taking place in their community or in communities across the country.

Another valuable source is USA.gov—the official site of the United States government. The website is directly linked to every federal and state agency. The information provided by the website is valuable to all citizens, and non-citizens, of the current news and regulations of the U.S. government. These are just some examples of e-government's influence in the Internet.

E-democracy leads to a more simplified process and access to government information for public-sector agencies and citizens. For example, the Indiana Bureau of Motor Vehicles simplified the process of certifying driver records to be admitted in county court proceedings. Indiana became the first state to allow government records to be digitally signed, legally certified and delivered electronically by using Electronic Postmark technology. In addition to its simplicity, e-democracy services can reduce costs. The Alabama Department of Conservation & Natural Resources, Wal-Mart and NIC developed an online hunting and fishing license service utilizing an existing computer to automate the licensing process. More than 140,000 licenses were purchased at Wal-Mart stores during the first hunting season, and the agency estimates it will save $200,000 annually from service.

Electronic democracy can also carry the benefit of reaching out to youth as a mechanism to increase youth voter turnout in elections and raising awareness amongst youth. With the consistent decline of voter turnout, e-democracy and electronic voting mechanisms can help revert that trend. Youth, in particular, have seen a significant drop in turnout in most industrialized nations, including Canada, the United States and the United Kingdom. The use of electronic political participation mechanisms may appear more familiar to youth and, as a result, garner more participation by youths who would otherwise find it inconvenient to vote using the more traditional methods. Electronic democracy can help improve democratic participation, reduce civic illiteracy and voter apathy and become a useful asset for political discussion, education, debate and participation.

Equally, these technologies are bellwethers of some of the issues associated with the territory, such as the inability to sustain new initiatives or protect against identity theft, information overload and vandalism.

Some traditional objections to direct democracy are argued to apply to e-democracy, such as the potential for governance to tend towards populism and demagoguery. More practical objections exist, not least in terms of the digital divide between those with access to the media of e-democracy (mobile phones and Internet connections) and those without, as well as the opportunity cost of expenditure on e-democracy innovations.

Furthermore, there are still those who are skeptical to the amount of impact that they can make through online participation. Although the government projects supply information, IT illiteracy and the digital divide are grounds to discourage participation. The political advances on the Internet can potentially dishearten non-users to adapt the new technologies.

ELECTRONIC DIRECT DEMOCRACY

Electronic direct democracy (EDD) is the strongest form of direct democracy, in which people are involved in the legislative function. The notion is utopian in the present capitalistic world, because the Internet and other electronic communications technologies are used only to ameliorate the bureaucracy involved with referendums. Many advocates think that also important to this notion are technological enhancements to the deliberative process. Electronic direct democracy is sometimes referred to by many other names.

EDD requires electronic voting or some way to register votes on issues electronically. As in any direct democracy, in an EDD, citizens would have the right to vote on legislation, author new legislation, and recall representatives (if any representatives are preserved).

Technology for supporting EDD has been researched and developed at the Florida Institute of Technology. The technology is currently used with student organizations.

EDD as a system is not fully implemented anywhere in the world although several initiatives are currently forming. Ross Perot was a prominent advocate of EDD when he advocated "electronic town halls" during his 1992 and 1996 Presidential campaigns in the United States. Switzerland, already partially governed by direct democracy, is making progress towards such a system. Senator On-Line, an Australian political party running for the Senate in the 2007 federal elections, proposed to institute an EDD system so that Australians can decide which way the senators vote on each and every bill. A similar initiative was formed 2002 in Sweden where the party Aktivdemokrati, running for the Swedish parliament, offers its members the power to decide the actions of the party over all or some areas of decision, or alternatively to use a proxy with immediate recall for one or several areas.

Liquid democracy, or direct democracy with delegable proxy, would allow citizens to choose a proxy to vote on their behalf while retaining the right to cast their own vote on legislation. The voting and the appointment of proxies could be done electronically. The proxies could even form proxy chains, in which if A appoints B and B appoints C, and neither A nor B vote on a proposed bill but C does, C's vote will count for all three of them. Citizens could also rank their proxies in order of preference, so that if their first choice proxy fails to vote, their vote can be cast by their second-choice proxy.

Digital Divide

The term digital divide refers to the gap between people with effective access to digital and information technology and those with very limited or no access at all. It includes the imbalances in physical access to technology as well as the imbalances in resources and skills needed to effectively participate as a digital citizen.

In other words, it is the unequal access by some members of society to information and communication technology, and the unequal acquisition of related skills. The term is closely related to the knowledge divide as the lack of technology causes lack of useful information and knowledge. The digital divide may be classified based on gender, income, and race groups, and by locations. The term global digital divide refers to differences in technology access between countries or the whole world.

The term initially referred to gaps in ownership of computers between groups, during which time the increase of ownership was limited to certain ethnic groups. The term came into regular usage in the mid-1990s, though the term had previously appeared in several news articles and political speeches as early as 1995. The President of the United States Bill Clinton and his Vice President Al Gore used the term in a 1996 speech in Knoxville, Tennessee. Larry Irving, a former United States head of the National Telecommunications Infrastructure Administration (NTIA) at the Department of Commerce, Assistant Secretary of Commerce and technology adviser to the Clinton Administration, noted that a series of NTIA surveys; were "catalysts for the popularity, ubiquity, and redefinition" of the term, and he used the term in a series of later reports. Since the start of the George W. Bush Administration, the NTIA reports have tended to focus less on gaps and divides and more on the steady growth of broadband access, especially amongst groups formerly believed to be on the wrong side of the digital divide.

There is a considerable literature on information and digital inequality that predates this current label. The concept of a digital divide is more of a new label and less of a unique concept.

There are various definitions of the term "digital divide". Bharat Mehra defines it simply as "the troubling gap between those who use computers and the Internet and those who do not".

This term initially referred to gaps in the ownership of, or regular access to, a computer. As Internet access came to be seen as a central aspect of computing, the term's usage shifted to encompass gaps in not just computers but also access to the Internet. Recently, some have used the term to refer to gaps in broadband network access. The term can mean not only unequal access to computer hardware, but also inequalities between groups of people in the ability to use information technology fully.

Due to the range of criteria which can be used to assess the imbalance, and the lack of detailed data on some aspects of technology usage, the exact nature of the digital divide is both contextual and debatable. Criteria often used to distinguish between the 'haves' and the 'have nots' of the digital divide tend to focus on access to hardware, access to the Internet, and details relating to both categories. Some scholars fear that these discussions might be discouraging the creation of Internet content that addresses these needs.

The discussions on digital divide often are tied with other concepts. Lisa Servon argued in 2002 that the digital divide "is a symptom of a larger and more complex problem the problem of persistent poverty and inequality". As described by Mehra), the four major components that contribute to the digital divide are "socioeconomic status, with income, educational level, and race among other factors associated with technological attainment".

Recognition of digital divide as an immense problem has led scholars, policy makers, and the public to understand the "potential of the Internet to improve everyday life for those on the margins of society and to achieve greater social equity and empowerment".

Typical measurements of inequality distribution used to describe the Digital Divide are the Lorenz Curve and Gini coefficient, however, the question of whether or not the digital divide is growing or closing is difficult to answer. In *Bridging the digital divide: An opportunity for growth for the 21st century*, examples of these ways of measuring are illustrated. In the Lorenz curve, perfect equality of Internet usage across nations is represented by a 45-degree diagonal line, which has a Gini coefficient of zero. Perfect inequality gives a Gini coefficient of one.

One area of significant focus was school computer access; in the 1990s, rich schools were much more likely to provide their students with regular computer access. In the late 1990s, rich schools were much more likely to have Internet access In the context of schools, which have consistently been involved in the discussion of the divide, current formulations of the divide focus more on how (and whether) students use computers, and less on whether there are computers or Internet connections. Public libraries and afterschool programs have also been shown to be important access and training locations for disadvantaged youth.

The E-Rate program in the United States (officially the Schools and Libraries Program of the Universal Service Fund), authorized in 1996 and implemented in 1997, directly addressed the technology gap between rich and poor schools by allocating money from telecommunications taxes to poor schools without technology resources. Though the program faced criticism and controversy in its methods of disbursement, E-Rate has been credited with increasing the overall number of public classrooms with Internet access from 14% in 1996 to 95% in 2005. Recently, discussions of a digital divide in school access have broadened to include technology related skills and training in addition to basic access to computers and Internet access.

Technology offers a unique opportunity to extend learning support beyond the classroom, something that has been difficult to do until now. "The variety of functions that the Internet can serve for the individual user makes it "unprecedentedly malleable" to the user's current needs and purposes".

Technology remains to be divided among SES in schools. The upper SES maintains access to technology at home, whereas, the lower SES children are limited to technology access only at school. With the non-equitable availability of technology outside of the classroom, there will continue to be a divide among student groups. Providing schools with technology is not sufficient to close the digital divide. Teachers must receive the appropriate training in order to use technology effectively and to increase student learning.

Although education could be used as a tool to close the "digital gap", closing this gap will not completely close the achievement gap among students from Lower SES backgrounds in comparison to Higher SES backgrounds.Another key dimension of the digital divide is the global digital divide, reflecting existing economic divisions in the world, which can clearly be seen in *The Global Digital Divide* image. This global digital divide widens the gap in economic divisions around the world. Countries with a wide availability of Internet access can advance the economics of that country on a local and global scale. In today's society, jobs and education are directly related to the Internet, in that the advantages that come from the Internet are so significant that neglecting them would leave a company vulnerable in a changing market. Andy Grove, the former chair of Intel, said that "[...]by the mid-2000s all companies will be Internet companies, or they won't be companies at all."

In countries where the Internet and other technologies are not accessible, education is suffering, and uneducated people and societies that are not benefiting from the information age, cannot be competitive in the global economy. This leads to these countries, which tend to be developing countries, suffering greater economic downfall and richer countries advancing their education and economy. However, when dealing with the global aspect of digital divide there are several factors that lead to digital divide. For example, country of residence, ethnicity, gender, age, educational attainment, and income levels are all factors of the global aspects of digital divide.In addition, a survey shows that in 15 Western European countries females, manual workers, elderly, and the less educated have less Internet access than males, professional, the young, and the well educated. The digital divide is a term used to refer to the gap between people who have access to the Internet and those that do not. It can also refer to the skills people have – the divide between peoples who are at ease using technology to access and analyse information and those who are not.

Digital Divide Worldwide

Canada

According to an Autumn 2001 Canadian Internet Use Survey, 73% of Canadians aged 16 and older went online in the 12 months prior to the survey, compared to 68% in 2005. In small towns and rural areas, only 65% of residences accessed the Internet, compared to 76% in urban areas. The digital divide still exists between the rich and the poor; 91% of people making more than $91,000/year regularly used the Internet, compared to 47% of people making less than $24,000. This gap has lowered slightly since 2005.

China

China is the largest developing country in the world and therefore saw their Internet population grow by 20% in 2006. However, just over 19% of Chinese people have access to the Internet and the digital divide is growing due to factors such as insufficient infrastructure and high online charges.

Europe

A European Union study from 2005 conducted in 14 European countries and focused on the issue of digital divide found that within the EU, the digital divide is primarily a matter of age and education. Among the young or educated the proportion of computer or Internet users is much higher than with the old or uneducated. Digital divide is also higher in rural areas. The study found that the presence of children in a household increases the chance of having a computer or Internet access, and that small businesses are catching up with larger enterprises when it comes to Internet access. The study also notes that "Despite increasing levels of ICT usage in all sections of society, the divide is not being bridged."

United States

According to a July 2008 Pew Internet & American Life report, "55% of adult Americans have broadband Internet connections at home, up from 47% who had high-speed access at home last year at this time [2007]". This increase of 8% compared to the previous year's increase of 5% suggests that the digital divide is decreasing. However, the findings go on to show that low-income Americans' broadband connections decreased by 3%. Therefore as was explained in the Digital Divide Evolution section, the detailed income groups need to be considered. Digital divide is a common subject in US politics and various government policies.

Digital Divide and Democratic Challenges

The theoretical concepts of e-democracy are still in early development, but many scholars agree that blogs (web logs), wikis and mailing lists may have significant effects in broadening the way democracy operates. There is no consensus yet among scholars about the possible outcomes of this revolution; it has so far shown promise in improving electoral administration and reducing fraud and disenfranchisement; particularly positive has been the reception of e-government services related to online delivery of government services, with portals (such as United States USA.gov in English and GobiernoUSA.gov in Spanish) used as intermediaries between the government and the citizen, replacing the need for people to queue in traditional offices.

One of the main problems associated with the digital divide as applied to a liberal democracy is the capacity to participate in the new public space, the cyberspace—as in the extreme case, exclusively computer-based democratic participation (deliberation forums, online voting, etc) could mean that no access meant no vote. Therefore, there is a risk that some social groups — those without adequate access to or knowledge of IT — will be under-represented (or others over-represented) in the policy formation processes and this would be incompatible with the equality principles of democracy.

Proponents of the open content, free software, and open access social movements believe that these movements help equalize access to digital tools and information.

Overcoming the Digital Divide

Projects like One Laptop per Child and 50x15 offer a partial solution to the global digital divide; these projects tend to rely heavily upon open standards and free open source software. The *OLPC XO-1* is an inexpensive laptop computer intended to be distributed to children in developing countries around the world, to provide them with access to knowledge. Programmer and free software advocate Richard Stallman has highlighted the importance of free software among groups concerned with the digital divide such as the World Summit on the Information Society.

Organizations such as Geekcorps, EduVision and Inveneo also help to overcome the digital divide. They often do so through the use of education systems that draw on information technology. The technology they employ often includes low-cost laptops/subnotebooks, handhelds (eg Simputer, E-slate,...), tablet PCs, Mini-ITX PCs and low-cost WiFi-extending technology as cantennas and WokFis. In addition, other information technology material usable in the classroom can also be made diy to lower expenses, including projectors.

Yet another solution is to try to better understand the lifestyle of a minority or marginalized community. In doing this, researchers can figure out "what is meaningful to them [minorities and marginalized users] and how they use (or do not use) different forms of the Internet for meeting their objectives". Furthermore, "a need for a re-examination of questions based on traditional ways of looking at people, their social dynamics, and their interactions with technology". However, researchers still tend to "set a 'method' for studying the impact of Internet use or assuming a golden rule for application that will function in all situations will not work". Additionally, "One strategy is to transfer goal-setting, decision making, and choice-determining processes into the hands of the disadvantaged users in order that they 'fit' Internet into their daily lives in ways that they themselves consider to be meaningful".

International cooperation between governments have begun, aiming at dealing with the global digital divide. For example, in an attempt to bridge this digital divide, an agreement between the United States Agency for International Development (USAID) and the Egyptian government emerged. The USAID funded state-of-the-art equipment for Egyptian education, their brilliance of knowledge in using such equipment caused such equipment to increase in use throughout the years. Now, Egyptian society is more computer literate and knowledgeable about computers than they used to be. Nonetheless it's a sign of progress that such attempts at bridging the digital divide are seriously being made. Additional participants in such endeavors include the United Nations Global Alliance for ICT and Development and the Digital alliance Foundation.

The United Nations is aiming to raise awareness of the divide by way of the World Information Society Day which takes place yearly on May 17.It also set up the Information and Communications Technology (ICT) Task Force in November 2001.

At the Massachusetts Institute of Technology, the IMARA organization (from Swahili word for "power") sponsors a variety of outreach programs which bridge the Global Digital Divide. Its aim is to find and implement long-term, sustainable solutions which will increase the availability of educational technology and resources to domestic and international communities. These projects are run under the aegis of the MIT Computer

Science and Artificial Intelligence Laboratory (CSAIL) and staffed by MIT volunteers who give training, installed and donated computer setups in greater Boston, Massachusetts, Kenya, Indian reservations the American Southwest such as the Navajo Nation, the Middle East, and Fiji Islands. The CommuniTech project strives to empower underserved communities through sustainable technology and education.

Some cities in the world have started programs to bridge the digital divide for their residents, school children, students, parents and the elderly. One such program, founded in 1996, was sponsored by the city of Boston and called the Boston Digital Bridge Foundation. It especially concentrates on school children and their parents, helping to make both equally and similarly knowledgeable about computers, using application programs, and navigating the Internet.

In the United States, minority ethnic groups have higher adoption rate for mobile communications devices than white Americans, to some degree leapfrogging over more expensive fixed-line Internet and PCs.

Each year, Certiport (which focuses on teaching digital literacy) awards the Champions of Digital Literacy award to leaders, world wide, who have helped to close the digital divide in their native countries..

The existence of a digital divide is not universally recognized. Compaine argues it is a *perceived* gap. Technology gaps are relatively transient; hence the digital divide should soon disappear in any case. The knowledge of computers will become less important as they get smarter and easier to use. In the future people will not need high-tech skills to access the Internet and participate in e-commerce or e-democracy. Thus Compaine argues that a digital divide "is not the issue to expend substantial amounts or funds nor political capital".

Citizens' E-participation

ICTs play a major role in organizing and informing citizens in various forms of civic engagement. ICTs are used to enhance active participation of citizens and to support the collaboration between actors for policy-making purposes within the political processes of all stages of governance. ICTs offer citizens not only the means to organize themselves, but also to produce cultural codes to represent themselves. ICTs can be seen as an important enabler to the empowerment of citizens or e-mancipation of citizens..

There has been much speculation about the Internet's potential to facilitate the engagement of younger citizens in politics. This group of young people, under the age of 35, frequently labelled Generations X and Y, have been noted for their lack of political interest and activity for the last two decades. The younger generation is less likely to have established long-standing habits of media use, and is willing to experiment with new technologies and formats. Young adults view the benefits of new technologies as a means of gaining advantage in education, employment, and in the political realm. Younger people have ease with Internet technology and are more likely than older citizens to use web-based platforms to research and access political information. However, there is no clear consensus about the capacity of new media, including the Internet and social networking websites, to engage young people in the democratic process.

E-participation is the generally accepted term referring to "ICT-supported participation in processes involved in government and governance". Processes may concern administration, service delivery, decision making and policy making. eParticipation is hence closely related to e-government and (e-)governance participation. The need for the term has emerged as citizen benefits and values have often received less attention in e-government development than those of the service providers, and the need to distinguish the roles of citizen and customer has become clearer.

A more detailed definition sees eParticipation as "the use of information and communication technologies to broaden and deepen political participation by enabling citizens to connect with one another and with their elected representatives". This definition includes all stakeholders in democratic decision-making processes and not only citizen related top-down government initiatives.

So eParticipation can be seen as part of eDemocracy, whereas eDemocracy means the use of ICT by governments in general used by elected officials, media, political parties and interest groups, civil society organizations, international governmental organizations, or citizens/voters within any of the political processes of states/regions, nations, and local and global communities. The complexity of eParticipation processes results from the large number of different participation areas, involved stakeholders, levels of engagement, and stages in policy making, which characterize the research and applications.

The term emanated in the early 2000s and draws generally on three developments.

1. The general development in CSCW (Computer Supported Cooperative Work) and groupware directed towards collaborative environments to support human ICT-mediated interaction, both work-related and social.
2. Developments in e-democracy since the late 1990s, where interest rapidly evolved from e-voting to several forms of ICT-supported and –enabled interaction between governments and citizens, including not only direct ones such as consultations, lobbying, petitioning and polling but also ones pursued outside of government itself, including electioneering, campaigning, and community informatics. To a large extent, the institutional framework conditions of the chosen democratic model define at which part of the democratic processes participation is permitted (such as direct or representative democracy, or any intermediate forms).
3. The development in e-government towards increasingly complex service delivery. Complex services require considerable interaction including searching, selecting options based on multiple criteria, calculating outcomes, notifications, inquiries, complaints, and many other activities. There are several ICT tools for such tasks, ranging from FAQs to call centres, but there is a need to coordinate all these into user-friendly but powerful toolsets for client-organization encounters. Because interaction in such contexts is complex, and because goals have to be reached, the arenas where it takes places becomes social arenas for ICT-supported participation.

The term participation means taking part in joint activities for the purpose of reaching a common goal. This encompasses both trivial situations in which participation mainly has a technical meaning, "doing things together". For example, a football team needs 11 players, and dancing often requires two or more people acting in a coordinated manner. But participation, even in trivial situations, also has a goal-oriented aspect which means decision making and control are involved. Participation in political science and theory of management refers to direct public participation in political, economical or management decisions. The two are not completely separated but belong on a spectrum of complexity and context.

When participation becomes complicated, decision making becomes necessary. Hence, any participatory process is potentially important for the rule system governing the activities. In terms of points 2 and 3 above, this means that when service processes become complex, the implementation of them will not be in all details based on political decisions but also on what is found to be practical. Such decisions are made in many places in a situated manner.

When such practical doings become implemented in government (e)service systems, they will affect decision making, as many changes will later be hard to make simply because exising procedures are implemented in ICT systems and government agencies' procedures. There are many theories dealing with institutionalization, for example Structuration theory, Institutional theory, and Actor-Network Theory. These theories all, in different ways, deal with how "ways of doing things" become established or rejected, and how those that become established increasingly affect the ways we "normally" do things.

Tools for e-Participation

A number of tools and models have emerged as part of the Web 2.0 that can be used or inspire the design of architecture for eParticipation. In particular, "the emergence of online communities oriented toward the creation of useful products suggests that it may be possible to design socially mediating technology that support public-government collaborations".

Participation Tools

1. Wikis
2. Online social networking
3. Blogs

Mechanisms

1. eVoting
2. Reputation systems
3. Internet petitions
4. Transparency tools

Tracking and Analysis

1. Digital traces

2. Data mining
3. Data visualization
4. Simulations such as agent-based social simulation

E-Participation Preparatory Action

eParticipation is the Preparatory Action lasts for three years (2006-2008). The EU is taking the lead in using on-line tools to improve the legislative process for its citizens. eParticipation which launched on 1 January 2007 will run as a series of linked projects which each contribute to a greater awareness and involvement by citizens in the legislation process right from the initial drafting through to its implementation at a regional and local level.

The individual projects will concentrate on ensuring that the legislative language and process is more transparent, understandable and accessible to the citizen. In addition the projects emphasis on the communication of legislation will be used to enhance and grow citizen's involvement and contribution in the process of creating and thereafter implementing the legislation.

So far, 21 projects have been funded. In the first year six trial projects were selected in 2007 seven more as well as the support action momentum to coordinate and promote the eParticipation projects and in 2008 a further seven trial projects were chosen. The European Parliament, national parliaments and local and regional authorities are actively involved.

State-of-the-art ICT tools are being tested to facilitate the writing of legal texts, including translation into different languages, and the drafting of amendments as well as making the texts easier for non-specialists to find and understand. New digital technologies are also being used to give citizens easier access to information and the more opportunity to try to influence decisions that affect their lives. A report which was published as a MOMENTUM whitepaper highlights the major facts and figures of those project's while providing some initial policy recommendations for future use. The European Commission has now launched a number of actions aiming at further advancing the work of supporting eParticipation.

Examples

1. FP7: ICT Challenge 7: Objective ICT-2009.7.3 ICT for Governance and Policy Modelling. The European Commission has launched some call in this area to finance researches.

2. CIP ICT Policy Support Programme (or ICT PSP). The European project has open a call in the programme CIP (Competitiveness and Innovation Framework) on the Theme 3: ICT for government and governance

Projects

a) **Demo-net**—The e Participation Network of Excellence. The overarching objective of DEMO-net is to strengthen scientific, technological and social research excellence in e-Participation by integrating the research capacities of individuals and organizations spread across Europe.

b) **Wave**—Welcoming Argument Visualization to Europe. The WAVE project "aims to improve the inclusiveness and transparency of EU decision making at the national and European level by using highly integrated, state-of-the-art Argument Visualization techniques to make the impact of complex EU environmental legislation on climate change more accessible and easy to understand for citizens, special interest groups and decision makers alike".

c) **Feed**—Federated e-Participation Systems for Cross-Societal Deliberation on Environmental and Energy Issues. The main objective Feed is to apply a new concept in e-Participation by allowing users to have seamless access to existing federated content that matches their needs for information on the several aspects of a debated issue as well as to informative material about the specificities of the deliberation procedure in the scope of which an issue is considered.

d) Demos Work—The objectives of Demos Work are to: address the issue of citizen disengagement from decision making processes, strengthen the EU fight against tobacco use, improve the quality of services provided to citizens, facilitate the work of MPs, accelerate transition to a knowledge society, increase the awareness of the Parliament activities among its citizens and leverage e-Representative assets.

Benefits of Online Participation

Several motivations lead people to contribute to virtual communities. Various online media (i.e. Blogs, Chat rooms, Electronic mailing lists, Internet forums, Wikis), are becoming ever greater knowledge-sharing resources. Many of these communities are highly cooperative and establish their own unique culture. They also involve significant time from contributors with no monetary gain.

Knowledge Sharing Infrastructures

Some key examples of online knowledge sharing infrastructures include the following:

1) Wikipedia: An online, publicly editable encyclopaedia with hundreds of thousands of editors
2) Slashdot: A popular technology-related forum, with articles and comments from readers. Slashdot subculture has become well-known in Internet circles. Users accumulate a "karma score" and volunteer moderators are selected from those with high scores.
3) Usenet: Established in 1980, as a "distributed Internet discussion system, it became the first medium for Internet communities. Volunteer moderators and votetakers contribute to the community.
4) Etc. (the Web2.0 is also referred to as the writable web for indicating that many people participate to the creation of its content)

In the past important online knowledge sharing infrastructures included:

a) AOL: The largest of the online service providers, with chat rooms which for years were voluntarily moderated by community leaders. It should be noted that rooms and most message boards are no longer moderated, however.
b) The well: A pioneering online community established in 1985. The WELL's culture has been the subject of several books and articles. Many users voluntarily contribute to community building and maintenance (e.g., as conference hosts).

Several researchers have investigated motivation in virtual communities. Studies show that over the long term users gain a greater insight into the material that is being discussed and a sense of connection to the world at large.

Peter Kollock researched motivations for contributing to online communities. Kollock outlines three motivations that do not rely on altruistic behavior on the part of the contributor: *anticipated reciprocity*; *increased recognition*; and *sense of efficacy*.

There is another motivation, implicit in the above, which Marc Smith mentions in his 1992 thesis: *Voices from the WELL: The Logic of the Virtual*

Commons: "*Communion*", as Smith terms it, or "sense of community" as it is referred to in social psychology.

Anticipated Reciprocity

A person is motivated to contribute valuable information to the group in the expectation that one will receive useful help and information in return. Indeed, there is evidence that active participants in online communities get more responses faster to Increased recognition.

Recognition is important to online contributors such that, in general, individuals want recognition for their contributions. Some have called this Egoboo. Kollock outlines the importance of reputation online: "Rheingold in his discussion of the Well (an early online community) lists the desire for prestige as one of the key motivations of individuals' contributions to the group. To the extent this is the concern of an individual, contributions will likely be increased to the degree that the contribution is visible to the community as a whole and to the extent there is some recognition of the person's contributions. ... the powerful effects of seemingly trivial markers of recognition (e.g. being designated as an "official helper") has been commented on in a number of online communities..."

One of the key ingredients of encouraging a reputation is to allow contributors to be known or not to be anonymous. The following example, from Meyers study of the computer underground illustrates the power of reputation. When involved in illegal activities, computer hackers must protect their personal identities with pseudonyms. If hackers use the same nicknames repeatedly, this can help the authorities to trace them. Nevertheless, hackers are reluctant to change their pseudonyms regularly because the status associated with a particular nickname would be lost.

On the importance of online identity: Profiles and reputation are clearly evident in online communities today. Amazon.com is a case in point, as all contributors are allowed to create profiles about themselves and as their contributions are measured by the community, their reputation increases. Myspace.com encourages elaborate profiles for members where they can share all kinds of information about themselves including what music they like, their heroes, etc. In addition to this, many communities give incentives for contributing. For example, many forums award Members points for posting. Members can spend these points in a virtual store. eBay is an example of an online community where reputation is very important because it is used to measure the trustworthiness of someone you potentially will do business with. With eBay, you have the opportunity to rate your experience with someone and they, likewise, can rate you. This has an effect on the reputation score. The participants may therefore be encouraged to manage their online identity in order to make a good impression on the other members of the community.

Sense of Efficacy

Individuals may contribute valuable information because the act results in a sense of efficacy, that is, a sense that they have had some effect on this environment. There is well-developed research literature that has shown how important a sense of efficacy is, and making regular and high quality contributions to the group can help individuals believe that they have an impact on the group and support their own self-image as an efficacious person.

Sense of Community

People, in general, are fairly social beings and it is motivating to many people to receive direct responses to their contributions. Most online communities enable this by allowing people to reply back to contributions (i.e. many Blogs allow comments from readers, one can reply back to forum posts, etc). Again, using Amazon.com as an example, other users can rate whether one's product review was helpful or not. Granted, there is some overlap between increasing reputation and gaining a sense of community. However, it seems safe to say that there are some overlapping areas between all four motivators.

In addition to participants that actively contribute to online discussions, many people join virtual community spaces and do not post, a concept referred to as lurking. There are several reasons why people choose not to participate online, and chief among them are: getting what they needed without having to participate actively, thinking that they were being helpful by not posting, wanting to learn more about the community before diving in, not being able to use the software because of poor usability and not liking the dynamics that they observed within the group.

Online Participation and Psychology

Studies have also found that the nature and the level of participation in online social networking sites have been correlated with the personality of the participants. The Department of Psychology in the University of Windsor found some levels of correlation in the articles "Personality and motivations associated with Facebook use" and "The Influence of Shyness on the Use of Facebook in an Undergraduate Sample". The articles state that people who have high levels of anxiety or stress or a shy personality are more likely to favor socializing through the Internet. The reason for this is because they are able to communicate with others without being face-to-face, and mediums such as chat rooms give a sense of anonymity, which makes them feel more comfortable in getting involved in discussions with their peers.

Information, Consultation and Participation

Over the last decade there has been a growing awareness of the need to develop new tools for public engagement that enable a wider audience to contribute to the policy debate and where contributions themselves are both broader and deeper. A number of commentators have addressed this issue and at the same time highlighted the possible dangers of a technology-driven approach.

Barber highlights the concept of strong democracy, creating active citizen participation where none had existed before. However he goes on to warn that the use of technology could diminish the sense of face-to-face confrontation and increase the dangers of elite manipulation. Held distinguishes nine different models of democracy. His participatory model reflects the need to engage both citizens and civil society organisations (CSOs) in the policy process. However, in order to engage citizens in policy-making, he and others recognise the need for informed and active citizens. Fishkin argues the need for "mass" deliberation by citizens instead of "elite" deliberation by elected representatives. Instant reactions to telephone surveys and television call-ins do not allow time to think through issues and hear the competing arguments. Fishkin states:

> A major part of the problem of democratic reform is how to promote mass deliberation – how to bring people into the process under conditions where they can be engaged to think seriously and fully about public issues.

Van Dijk addresses the role of information and communication technology with such participatory models of democracy in order to inform and activate the citizenry. However he warns of the consequences of poorly designed technology.

Computerized information campaigns and mass public information systems have to be designed and supported in such a way that they help to narrow the gap between the "information rich" and "information poor" otherwise the spontaneous development of ICT will widen it.

Several commentators discuss the broader use of technology to support the democratic process. Coleman and Gøtze outline four possible scenarios for technology supporting democracy. The first e-democracy model is where the technology supports direct democracy. For example, Becker and

Slaton explore the current state and future of e-democracy initiatives that are designed specifically to move towards direct democracy.

The second model is based on online communities, where technology is concerned with supporting civic communities. The work of Rheingold on virtual communities assesses the potential impact of civic networks, questioning the relationship between virtual communities and the revitalisation of democracy. Tsagarousianou *et al.,* give descriptions of a number of projects involved with e-democracy and civic networking.

Coleman and Gøtze's third e-democracy model concerns the use of online techniques to gauge public opinion through surveys and opinion polls. However, Fishkin questions whether opinion polls contribute to the complex issues of public policy. He argues that as far as American citizens are concerned they have the opportunity to be consulted on several occasions by opinion polls without prior warning or preparation, in order to find their views, even when the individual may have had no reason to develop any opinion on the subject being asked. He concludes that all that was gathered was an "attitude" created on the spot by the very process of participating in the survey.

Finally, their fourth model focuses on the use of technology to engage citizens in policy deliberation, emphasising the deliberative element within democracy. Fishkin sets out some general conditions for deliberation concluding that deliberation requires:

> Conditions that reknit the citizenry to the political process: that encourage thoughtful discussion, mutual respect, active participation and an openness of the process to all groups and strata. We must create public spaces that effectively motivate citizens to become a "public" where realisation of these values is possible.

Coleman and Gøtze build on these requirements and define deliberative engagement as:

> Methods of public engagement can be described as deliberative when they encourage citizens to scrutinize, discuss and weigh up competing values and policy options. Such methods encourage preference formation rather than simple preference assertion.

Increasingly, OECD member and non-member country governments around the world are committed to facilitating wider public participation in policy-making. Typically they are addressing issues such as how to provide easier and wider access to government information and how to ensure that citizens have the ability to give their views on a range of policy related matters. This top-down perspective of democracy is characterised in terms of user access to information and reaction to government led initiatives. Dutton focuses on technology to support access to politics and services with the citizen as the user. Sclove argues that this top-down approach alone is inadequate for strengthening democracy, stating:

> Numerous political theorists agree that decision-making processes are democratically inadequate, even spurious, unless they are combined with relatively equal and extensive opportunities for citizens, communities, and groups to help shape decision-making agendas.

From the bottom-up perspective, citizens are emerging as producers, rather than just consumers, of policy. Recognising that there is a need to allow citizens to influence and participate in policy formulation, technology is emerging to support this. However, most citizens have not been involved typically in pre-policy planning. Instead of reacting to an agenda set by government, this bottom-up perspective allows CSOs and citizens to set the agenda.

The OECD report, *Citizens as Partners: Information, Consultation and Public Participation in Policy-making,* discusses the government-citizen relationship in policy-making and argues that the bringing together of the top-down and bottom-up approaches can enable a strong partnership leading to a strengthening of representative democracy. The document defines three types of interaction, namely one-way information provision, a two-way consultation relationship where citizens are given the opportunity to give feedback on issues and, lastly, active participation – a relationship based on partnership where citizens are actively engaged in the policy-making process. The report highlights 10 guiding principles for successful information, consultation and participation in the policy-making life cycle.

The Report of the Government Commission on Swedish Democracy looks beyond traditional consultation mechanisms and emphasises the need to increase citizens' participation and influence on the development of society in the 21st century. With regard to technology it states:

> IT can also be used to reinforce civil society and to promote participative democracy. The new information technology has the potential to broaden opportunities for citizens to participate in and influence problem formulations and discussion before decisions are made by elected assemblies. National and local policy should in the first instance be directed at developing techniques and methods for such participative democracy with the support of I T.

In November 2001 the UK government published a report on "Better Policy-Making" based on a survey of senior policy-makers in the UK. This looked at the barriers and enablers and identified where the policy-making process needs to change to enable policy-makers to move into the 21st century. It states that modern policy-making should contain the following nine features: be forward looking, be outward looking, be innovative, flexible and creative, ensure decisions are based on the best available evidence (*i.e.* be evidence-based), be inclusive, be joined up, ensure policy is constantly reviewed, enable systematic evaluation, and learn from experience.

Enacting democratic principles, building the enabling infrastructures, and engaging citizens in political decision-making is expected to lead to more active citizenship and good governance, and has resulted in a large number of democracy pilot studies on information dissemination, consultation and participation in policy-making. The OECD report *Citizens as Partners, Information, Consultation and Public Participation in Policy-making* documents a number of these initiatives and details specific case studies in some of OECD member countries.

There are also a growing number of examples of government organisations innovatively using ICTs, and in particular the Internet, to provide access to policy information and request public comments on it. These examples demonstrate how technology is emerging as a tool not just to disseminate information but also to provide people with the capacity to participate and influence decision-making. The report "Online Consultation in GOL Countries", which is a joint product of the members of the Government Online International Network, provides a good description of the current state of using technology to inform and engage citizens.

In July 2002 the UK government issued a consultation paper on a policy for electronic democracy. This consultation document usefully argues that e-democracy can be divided into two distinct areas - one addressing e-engagement and the other addressing e-voting. In the case of the former, the document sets out the possibilities for greater opportunity for consultation and dialogue between government and citizens.

Given the expanding knowledge base of e-engagement practice and the emergence of government e-engagement policy, there is every indication that the use of ICT to engage citizens will increase. However there is, at present, no methodological framework that specifically addresses how ICT can be designed and used to efficiently and effectively support information provision, consultation and participation in the policy-making life cycle. Based on the input and deliberations of the OECD member countries participating in the OECD Expert Group on Government Relations with Citizens and Civil Society, this report seeks to contribute to the development of such a framework.

Objectives of E-engagement

The OECD report *Citizens as Partners* argues that democratic political participation must involve the means to be informed, the mechanisms to take part in the decision-making and the ability to contribute and influence the policy agenda. Specifically it usefully defines the following terms).

- *Information*: a one-way relation in which government produces and delivers information for use by citizens. It covers both "passive" access to information upon demand by citizens and "active" measures by government to disseminate information to citizens.
- *Consultation*: a two-way relation in which citizens provide feedback to government. It is based on the prior definition by government of the issue on which citizens' views are being sought and requires the provision of information.
- *Active participation*: a relation based on partnership with government, in which citizens actively engage in the policy-making process. It acknowledges a role for citizens in proposing policy options and shaping the policy dialogue - although the responsibility for the final decision or policy formulation rests with government.

ICTs have the potential to support the development of e-engagement applications that address these joint perspectives of informing, consulting and participating. A growing number of publications reflect the growing maturity of electronic democracy as a field of research and practice.

One can consider three over-arching reasons for better engagement of citizens in the policy-making process:

- To produce better quality policy.
- To build trust and gain acceptance of policy.
- To share responsibility for policy-making.

Other reasons include, but are not limited to, the ability to visualize policy, to give citizens the means to supervise government and policy implementation, to balance the power of lobby organisations, to avoid corruption, to foster active citizenship, all with the long term objective of strengthening representative democracy.

Given these overarching reasons, the objective of technology-enabled information dissemination, consultation and participation is to improve the policy-making process through a range of devices designed to enable:

- Reaching and engaging with a wider audience through a range of consultation and participation technologies adapted to cater for the diverse technical and communicative skills of citizens thereby enabling broader participation.
- Providing relevant information in a format that is both more accessible and more understandable to the target audience to enable more informed participation.
- Enabling more in-depth consultation and supporting deliberative debate online.
- Facilitating the analysis of contributions to support policy-makers and to improve policy.
- Providing relevant and appropriate feedback to citizens to ensure openness and transparency in the policy-making process.
- Monitoring and evaluating the process to ensure continuous improvement.

Additional benefits include the possibility to build seamless government, to facilitate information storage and retrieval, and to enhance the attractiveness of a debate for certain audiences.

It is also necessary to determine where in the policy-making process technology-enabled information dissemination, consultation and participation can be effective. The OECD report *Citizens as Partners* states that:

> "Citizens can make an active and original contribution to policy-making when their relationship with government is founded on the principle of partnership. Active participation represents a new frontier in government-citizen relations for all OECD countries and different countries use different terms ... it requires governments to share in agenda-setting and requires commitment from government that policy proposals generated jointly will be taken into account in reaching a final decision."

The policy-making process can be considered as a cycle of activities which include preparation, implementation and evaluation and where results are fed back into agenda-setting and policy design. The report provides a table indicating information, consultation and participation in the policy cycle.

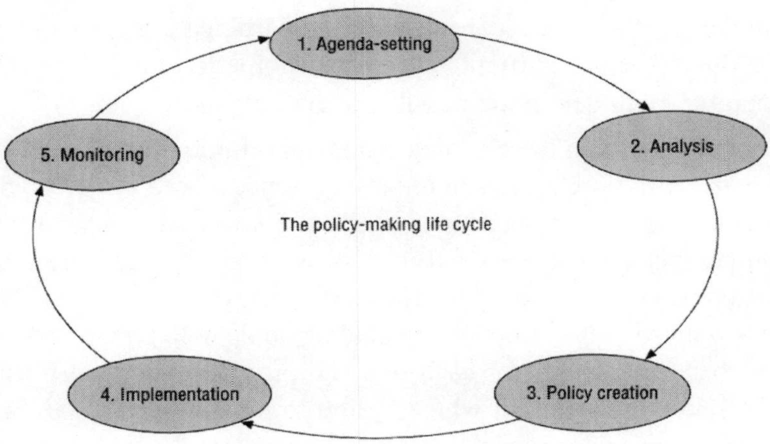

Figure 1. The policy life cycle

In order to establish a framework for discussing where ICT is most appropriate in the policy process, and to ensure that all readers are using the same terminology, the following five stages for policy-making are suggested (Figure 1). These are based on discussions with representatives from the UK Cabinet Office and the Office of the E-Envoy. Clearly, there will be differences across the OECD member countries, with countries using different terminology and possibly not including all the stages in their current policy-process, but by defining these stages and with these provisos, it will be possible for countries to jointly discuss the design and appropriateness of technology throughout the policy-making process and so share experiences.

These stages will act as a baseline for developing a framework for e-engagement in the later sections. It is the intention to look at each of them and determine how information, consultation and participation can be supported at each stage through the use of appropriate technology. Each country, of course, has the possibility to develop a more detailed set of stages for their specific policy-making process.

1. *Agenda setting*: establishing the need for a policy or a change in policy and defining what the problem to be addressed is. This may arise as the result of a change of government; a sudden change in the environment; a growing development; a new problem or a continuing problem. Information, consultation and participation are all important for this stage. In particular active participation allows citizens the opportunity to determine possible agenda items.
2. *Analysis*: defining the challenges and opportunities associated with an agenda item more clearly in order to produce a draft policy document. This can include: gathering evidence and knowledge from a range of sources including citizens and civil society organisations; understanding the context, including the political context for the agenda item; developing a range of options (including doing nothing) and conducting cost benefit analysis for each one and providing advice to Ministers who take a decision on which option to pursue. Again, information, consultation and participation are all important for this stage. Active participation allows citizens to determine the range of options under review.
3. *Creating the policy*: ensuring a good workable policy document. This involves a variety of mechanisms which can include: formal consultation, risk analysis, undertaking pilot studies, and designing the implementation plan. Ministers take decisions. At this stage information and consultation are important but there is possibly limited scope for active participation.

4. Implementing the policy: this can involve the development of legislation, regulations, guidance, and a delivery plan. At this stage information and consultation are important but there is also scope for active participation (*e.g.* in the delivery of public goods and services by civil society organisations).
5. *Monitoring the policy*: this can involve evaluation and review of the policy in action, research evidence, views of users and horizon scanning. Information, consultation and participation are all important for this stage. Active participation allows citizens the opportunity to give their views on the policy in action and to suggest changes.

At this point the process can loop back into stage 1, as the policy may be modified on the basis of experience with implementation.

OECD guiding principle number 4 with regard to the issue of time in the policy life cycle, states:

> Public consultation and active participation should be undertaken as early in the policy process as possible to allow a greater range of policy solution to emerge and to raise the chances of successful implementation. Adequate time must be available for consultation and participation to be effective. Information is needed at all stages of the policy cycle.

Currently, the UK government generally initiates consultation with "experts" that is academics, delivery staff, businesses and other NGOs at stages 1 and 2. At stage 3 this consultation is broadened out to include citizens. Stage 5 has traditionally been handled through formal evaluation of the policy and the use of management information.

Each of the above policy-making stages has the potential to be strengthened by the use of ICTs to better disseminate information leading to a more informed citizenry, to enable better consultation helping to secure broader and more informed contributions, and to support better participation facilitating the emergence and articulation of bottom-up ideas influencing policy formulation and agenda-setting. We will return to this framework and develop it further based on issues of the appropriateness of technology discussed in the following sections.

Some OECD member countries have developed guidelines for online consultations, for example, Canada and the Netherlands. However it is not obvious from these guidelines how the contributions from the citizens will be integrated into the government process of policy-making. It is unclear what, if any, mechanisms are in place to ensure that contributions are incorporated appropriately and that such contributions actually have an effect on the policy content. Indeed many reports on e-democracy report on pilots at national and local level of government. However the majority of these reports only focus on specific aspects of engagement in the policy-making process, that is, the "front-end" aspects where information is disseminated and where comments are sought. The corresponding "back-end" processes, that is, where the input from citizens has to be analysed, integrated into policy and feedback provided to those who contributed, have less weight placed on them or are not discussed at all.

Most commentators focus on the ability of governments to improve the dissemination of information to citizens using technology and are silent as to how technology could support the better production of information in the first place. Others focus on how technology can better gather contributions from citizens but typically overlook how technology could support the analysis and feedback of comments to citizens. ICTs need to be designed to support the whole of the policy-making life cycle, both the front-end processes involving the engagement of citizens and the back-end processes involving the integration of citizens' contributions into policy. This holistic view of technology-enabled policy-making will be re-visited in the following sections.

Design of E-engagement Systems

This section considers the issues and constraints that have to be taken into account when designing ICT based tools to enable citizen engagement in the policy process. We specifically focus on the general design issues for e-engagement systems. We consider who the "users" are and how, if possible, to involve these users in the design of the systems themselves.

Because we are considering the use of technology to support democratic practice, design has to be considered from the two perspectives of "technology" and "democracy". We take as our starting point in the design of e-engagement systems relevant existing guidelines on best practice. From the democratic perspective we use existing guidelines for citizen engagement and from the technology perspective we use software best practice for participatory user design.

By considering the objectives of technology-enabled information dissemination, consultation and participation as discussed in the last section we can begin to draw up some guiding principles that are relevant to all electronic engagement systems. Design aspects that are specific to the separate components are discussed in the following sections on technology to support information, consultation and participation.

Reaching and engaging with a wide and diverse audience highlights the importance of designing systems that provide the widest possible accessibility and ease of use. Accessibility, usability and security are widely considered to be critical issues in the design of systems and services for the general public. The contentious nature of governance means that, in the design of e-engagement systems, these issues can become more complex. Democratic needs for openness and transparency may conflict with needs for ease of use and simplicity of access. The issues of unequal access to technology and the unequal technical capabilities of citizens demand systems that are simple to use. Similarly the demand for transparency may call for procedures to be streamlined and simplified. However when procedures are questioned or fail, demands can be made for explanations of why procedures are as they are, and their wider legitimacy may be called into question.

For e-Government, trust refers to trust in both government and the technology. Establishing trust demands both that government can establish that those who seek access to information, and decision-making have a legitimate right to do so, and that the privacy and civil rights of those who are governed are not compromised in doing so. This issue of trust in government is a significant area for research. The experience of electronic commerce has emphasised contractual trust, and focused on technological issues like security and authentication. Trust is an important resource in the e-commerce environment and shared knowledge of the rules, which shape its generation, maintenance, and impacts are critical to the success of e-commerce transactions. This is equally true for governments, if not more so. However, in addition representations of trust, trustingness and trustworthiness which take a more socially-orientated approach are required for government initiated e-engagement systems.

The following design issues and trade-offs need to be explored:

- Balancing the need for straightforward, anonymous access to systems, with the need to collect personal data for various reasons such as authentication and evaluation.
- Balancing the needs for standard, generic interface features with the need to reflect the expectations of a variety of target audiences.

- How to support easy and flexible navigation through complex policy issues.
- Deciding how much information should be provided to assist individuals to be adequately informed on issues and so have the competences to contribute.
- Finally, balancing rights of access, protection of privacy and security with issues of transparency, accountability and trust.

One clear guiding principle from the democracy perspective is the need to state objectives clearly. The OECD guiding principle number 3 on "clarity" states:

> Objectives for, and limits to, information, consultation and active participation during policy-making should be well-defined from the outset. The respective roles and responsibilities of citizens (in providing input) and government (for making decision for which they are accountable) must be clear to all.

The e-consultation guidelines published by the Netherlands government advises that it should "be made particularly clear in advance what is expected of the participating citizen and within what scope he is to operate".

Having clear objectives helps to identify and scope the target audience. There is a need to understand who the audience is, what their communicative and technical skills are likely to be, and importantly, what will motivate them to become engaged in the policy-making process. Assuming all citizens to be the target audience immediately causes difficulties because of the diverse nature of such a large audience. Designers suggest that audiences be narrowed down as far as possible in order that the "look and feel" of the e-engagement system can be matched to the cultural attitudes of the audience. E-engagement systems for young people are more likely to attract them if the design meets the goals and expectations of that age group. In order to achieve an acceptable system it is important to involve a cross-section of the target audience in the design of the system as early as possible.

The typically diverse nature of the target audience has technical implications as far as accessibility is concerned. At this point in time most Internet-based engagement pilots have had to be designed to reflect the lowest denominator of requirements. It is important that the e-engagement system can be used by people using any Internet enabled computer at minimum connection rates and with any browsers. To achieve this, the majority of current e-engagement pilots are designed to be predominantly text based with a few light graphics. The use of video, audio or animation techniques has to be carefully thought through because of the above reasons.

Usability and clear navigation for Internet-based systems are important aspects of design. Powazek argues that clear navigation is one way to help achieve a user's trust in the system. For Internet-based systems, accessibility and usability features for partially sighted users and blind users need to be addressed. For example, all images should given tags so that a textual description of their function will be available in their place for people accessing the site with a screen reader. Given the increased access and use of the Internet these issues will become more important. This is discussed in detail and recommendations have been made by the World Wide Web Consortium through their Web Accessibility initiative.

In the US, specific legislation (Section 508 of the Rehabilitation Act) requires any electronic information developed, procured, maintained, or used by the federal government to be accessible to people with disabilities. All electronic information created or acquired by any federal agency or department must also comply with these accessibility standards.

Control mechanisms, such as user registration and issuing user IDs and passwords need to be considered. However the extent to which they are required can be gauged by asking the questions such as: "Are the results of the engagement action legally binding to government or rather informing Ministers on the issue?", "Is the information of a confidential nature and therefore can only be made available to a restricted audience?"

For online consultation and participation through discussion forums clear engagement guidelines need to be issued. These should contain a Privacy Policy, a statement on whether Anonymous Comments are allowed and Conditions of Use. The important need for moderation and making rules clear from the start is stressed in Coleman and Gøtze.

An example of a condition of use statement is:

> "We reserve the right to delete comments that may, in our view, be considered libellous. Users wishing to make comments on the consultation document are requested to refrain from using offensive or abusive language, to refrain from including advertising statements or including text of a disruptive nature. Users should be aware that any such comments may be removed. Inclusion of any statement or comment in this site does not indicate that consultation sponsors endorse it or take any responsibility for it."

E-engagement systems are relatively new therefore one can expect the design of such systems to evolve over time. To support on-going improvement in these systems it is imperative that the design of the system include an opportunity for users to comment on the appropriateness of the user interface, content and functionality of the system. One way of achieving this is to include an online questionnaire or assessment form for users to complete.

Promote your Online Consultation

Promoting e-engagement is an important issue. We need to find new ways to promote online information, consultation and participation to citizens. One of the objectives of e-consultation is to reach a wider audience, and not just the organisations and individuals that usually respond to consultations. To achieve this objective the methods for promoting the online consultation have to be broader than for traditional consultation exercises. A comprehensive publicity plan needs to be drawn up and followed right from the start of the e-engagement exercise. This plan could include:

- A high profile launch of the e-consultation, remembering to include the Web address in any press release.
- Availability of Internet connected PCs at related events for citizens to make contributions along with demonstrations on how to use/contribute to the e-consultation.
- Web links from stakeholders and government websites.
- Invitations to politicians to actively contribute online.
- Paper postcards promoting the site and inviting comments.
- "Tell a friend" type e-postcards.
- Clickable banner ads on the websites most visited by the target audience should be considered and allowed for in the e-consultation budget.

In conclusion, among the lessons learnt from previously reported e-engagement studies is to enlist the support of stakeholders who represent a cross section of the target audience when designing the systems. E-engagement is not radically different from traditional citizen engagement, therefore it is important to build on both best practice for consultation and best practice for designing systems for a diverse audience.

Information Online

Information: is a one-way relationship in which government produces and delivers information for use by citizens. It covers both "passive" access to information upon demand by citizens and "active" measures by government to disseminate information to citizens.

One of the objectives of technology-enabled information dissemination is to provide relevant information in a format that is both more accessible and more understandable to the target audience to enable more informed participation and acceptance of policy. These two aspects of accessibility and understandability will be addressed in this section.

As reported elsewhere many governments are providing Web portals for citizens to gain access to information. It is also worth noting that parliaments in many OECD member countries are also providing extensive websites for citizens. We will not address the issue of how these government and parliament websites should or should not be linked, but leave readers to contemplate this question.

One example of a parliament providing extensive information via the Web is the Scottish Parliament's website. This site publishes the Official Report of the Parliament's meetings in the Chamber by 7 a.m. on the following day, and Committee Reports as soon as possible and generally within 3 days of the meeting. In addition, and as a further step towards openness, committee agendas and papers are published in advance where possible. The site also shows the email addresses of all MSPs, their biographies and links to personal Web pages, alongside phone and fax numbers and correspondence addresses. Email addresses for clerks to committees and for offices within the Parliament are also published. A webcasting service broadcasts Chamber and Committee meetings live across the Web so that organisations and individuals can listen in and watch items of particular importance to them. The service also includes access to the audiovisual archive of webcasts, and relevant papers for the committee meetings.

It is also worth noting that in the US, the "Congress online" project has considered how the various websites on Capital Hill should be organised and a report has been published. It discusses the critical elements for building and maintaining effective congressional websites and provides the results of their evaluation.

Access to Information

Access to information is the cornerstone on which citizen engagement rests and requires both legislation and clear government mechanisms for its enactment. There is a need to reconcile the citizens' right to know with the individual's right for privacy and the need to preserve confidentiality, where disclosure of information would be against the pubic interest. Once these issues are addressed, how can technology support access to information?

The Government Online report on electronic consultation distinguishes between the terms access and accessibility.

- Access means the real possibility of consulting or acquiring government information electronically.
- Accessibility means the ease with which one can actually make use of the possibility of consulting government information electronically.

The report recognises seven characterises that determine the degree of accessibility which are very relevant here. They are:

1. *Recognisability and localisability*: the public must know what information is available from which government body, and how and where this can be located.
2. *Availability:* the information must be stored in a standard digital form and be accessible through an electronic medium or data carrier.
3. *Manageability*: the public must not drown in the quantity or complexity of the information but instead be able to find their own way through the system, if necessary using search systems provided by the government.
4. *Affordability:* the price of the information should not create any barriers, this being dependent on the importance of the information for society at large.
5. *Reliability:* the public must be able to rely on the correctness, completeness and authenticity of the information.
6. *Clarity:* the information must be as clear as possible in terms of content, context and presentation.
7. *Special needs:* the information should preferably also be accessible to the blind, the visually handicapped and people with other handicaps.

These characteristics usefully focus on access in terms of the appropriate form of the information and its appropriate location given the diverse nature of the potential audience. Citizens may wish to access information over the Internet using number of devices and from a range of locations. These include public access points in libraries and community cyber cafes, kiosks, schools and also access from the workplace and the home.

It is not necessarily easy for citizens to know where relevant information is located. Search engines can facilitate this but these typically result in long lists of, often, irrelevant information. The development of government portals is helping to address this problem to some extent. Some government sites.

Understanding Information

Due to the expected diversity of users, the aim should be, wherever possible, for text to be written clearly and simply with the avoidance of jargon and legalistic terms. In addition, governments in several OECD member countries must respond to legal obligations to publish information in all official languages. Even in the absence of such legal requirements, many governments provide information in the languages of important ethnic minorities. How can technology support this?

Support for such efforts may emerge from advances in research on Natural Language Processing (NLP). Work on multi-lingual access and style checking, which can be considered *language-oriented* research and development, has begun to address the problems related to understandability. Such NLP tools can be used to check the style of documents according to standard readability measures and to mark up important and difficult expressions. Two specific types of support are discussed here, namely support for style checking and support for language translation.

Style

Style checking of a document identifies "strange" words and expressions. One way of achieving this is to link legalistic and technical expressions to an online help menu or glossary. In this way, the user can navigate through the information and decide which words he or she needs an explanation for. Early experiments in this field came from applied research in psycholinguistics. Here the focus was on readability and comprehensibility. For a good review of these studies see Lehner. To check the style of writing many different techniques have been developed which process documents in order to recognise some mathematical features of the text. These techniques are generally referred to as *readability* or *comprehensibility* formulae. Although this kind of text processing does not ensure, from a theoretical point of view, a reliable evaluation of text, readability formulae have been successfully applied to several domains in order to ensure clarity and ease of reading of public information. Current research in this field is generally referred to as *controlled writing*. These techniques are being applied to many different domains, such as technical documentation and legal documentation, *i.e.* domains where it is crucial to ensure broad and easy comprehension of the content of a document. In many cases, readability evaluation techniques are associated with static resources such as guidelines or glossaries, and can be used as optional tools by the writer.

Multi-lingual translation

Multi-lingual translation is important when information is to be disseminated and read by all citizens in countries with more than one official language or coming from different ethnic backgrounds. However the field of automatic translation is recognised as a difficult research area. The UNL project is an important current research project in this area. The website describes the project:

> UNL (Universal Networking Language) is a language that – with its companion "enconverter" and "deconverter" software – enables communication among people with differing native languages. It will reside, as a plug-in for popular World Wide Web browsers, on the Internet, and will be compatible with standard network servers. The technology will be shared among the member states of the United Nations. Any person with access to the Internet will be able to "enconvert" text from any native language of a member state into UNL. Just as easily, any UNL text can be "deconverted" from UNL into native languages.

Popular examples of translation systems are typically simple "assistants" using online support such as thesauruses and dictionaries. They tend to be relatively simple systems and because of this have difficulty dealing with ambiguous or complex terms.

Whether assisted by new ICTs, or simply as the result of careful drafting, the overarching aim is to ensure that citizens have access to relevant, complete, timely and understandable information on the basis of which they may then formulate their own opinions and proposals for policy.

E-CONSULTATION

Consultation is a two-way relationship in which citizens provide feedback to government. It is based on the prior definition of the issue on which citizens' views are being sought and requires the provision of information. Governments define the issues for consultation, set the questions and manage the process, while citizens are invited to contribute their views and opinions.

ICT provides the potential to allow policy-makers to go directly to users of services and those at whom a policy is aimed to seek their input. While government would always have to weigh up carefully the different types of evidence available regarding the likely success of its draft policy, this is a potentially valuable addition to the range of evidence available. When consultation is merely informing government and where government reserves the right to make the final decisions, it can be argued that citizens will be able to have greater influence on policy content through consultation earlier in the policy making process rather than later. It can also be argued that consultation at the stage of a draft policy document requires citizens to have the communication skills to interpret the typical legalistic terminology of the document before commenting appropriately. Whereas if citizens are given the opportunity to comment before this stage in policy-making they will still need to be well-informed on issues, but the information could be made more readable and understandable. This could be achieved using technologies discussed in the previous section.

These points are highlighted by Rosen in a review of e-democracy trials by Swedish local authorities. He states that consultation is usually carried out at the policy creation stage, when:

> It is unclear what the final decision may be, which in turn means that consultation may also be needed at a later stage when the final decision has begun to take shape.

This finding underlines the need for consultation at several stages of the policy process. On the other hand, the Netherlands government e-consultation guidelines, argue for consultation earlier rather than later, stating that e-consultation:

> Will only make sense if the process of making decisions on a certain subject is in an early phase; in other words, when the matter is still open to modification.

A Sesign Template for Online Consultation Websites

Overview: a welcoming page outlining the purpose, target audience, timetable of the consultation, who is undertaking the consultation, and why. It may include a clear statement regarding the conditions of use of the website (*e.g.* specifying that offensive language will be removed). In general, the overview should follow best practice guidelines for off-line consultations as referenced elsewhere in this book.

Background information: this is where citizens can be informed about the consultation issue. It can comprise comprehensive pages on the consultation subject or link to other electronic sources for more detailed information. The accessibility and understandability of such information, as discussed in the previous section, is crucial.

Consultation tool/technique: the exact tool/technique used to gather citizens' comments depends on the type of consultation being undertaken. Several of the examples given above could fit into this framework, *e.g.* online surveys, opinion polls, discussion forums.

Other ways to be involved: this is where details of events and other nonelectronic consultation activities associated with the topic are described, providing a level of co-ordination and integration between off-line and online consultation exercises.

Feedback: space for a statement from those organising the consultation on the results and effects of the consultation once the consultation is complete and contributions have been analysed.

Review site: an online questionnaire for users to complete in order to support continuous improvement of the consultation exercise.

Contributors: a list of the names and countries of all those who make a contribution can be displayed here, subject to national data protection legislation.

Log-in/Logoff: this is where user registration occurs, if necessary.

Tell a friend: this allows the promotion of the online consultation electronically, for example, users can automatically email people they feel would be interested in participating in the consultation.

The above is provided as a template of what could be appropriate on e-consultation Web pages. The actual consultation technique can be slotted in as appropriate or multiple consultation techniques used. Whichever consultation technique is used, most commentators agree for the need to encourage and support informed participation by the citizens. For discussion forums, Powazek suggests burying the "post it" button in order to deter "spur of the moment" responses, noting that: "The more clicks it takes, the better the post will be."

As well as the citizen-orientated sections, the management of the e-consultation process could be facilitated by additional password-protected administrative services. These could include functions to:

- View and monitor the comments added in the last 24 hours.
- Remove from view comments that breached the "conditions of use" statement.
- View the most frequently read comments.
- View the comments received from specific postcode areas.
- View the entries to online user-evaluation questionnaire.

Again, this is only a sample of possible administrative facilities that are needed and therefore could be included in any e-consultation system. By providing the appropriate "administrative" functions, policy-makers have the ability to read comments and, relatively easily, judge the progress of the e-consultation right from the start of the exercise.

Technology to Support Online Deliberation

In the 1970s Horst Rittel termed problems characterised by indeterminacy and complexity as "wicked" problems. These types of problems require deliberative discussion where consensus can emerge through debate with alternative options and competing interests being exposed. The policy-making process is a good example of this type of "wicked" problem and can typically be characterised as:

- Not easily defined in such a way that all stakeholders agree that this is the problem to solve.
- Has no clear stopping rules, it is only the time limits on consultation that mark the end of a problem solving period.
- There is no clear cut right or wrong approach, instead there are better or worse solutions.
- Difficult to provide objective measures of success.

- Policy making is an iterative process.
- Solutions and/or options have to be discovered.
- Although policy options may be similar, each is unique.
- The level of detail necessary to define the problem is a matter of judgement.
- There is strong moral and political pressure against failure.

To support the solution of these types of problems Rittel developed IBIS (Issue-Based Information Systems) – a language and graphical representation scheme for visualising argumentation. It was originally a hypertext environment for the structured discussion of design issues.

Computer Supported Collaborative Work (CSCW) and Computer Supported Collaborative Argumentation (CSCA) – the latter developing since the 1970s – are two major areas of research that provide support for e-consultation. Two examples of projects are provided below to give an indication of the type and extent of research in this area and its relevance to e-consultation.

Online Moderation Systems

The ZENO system was originally developed from work in CSCA. Zeno is a mediation system, *i.e.* a discussion forum with special support for deliberation, the process of discussing, arguing and negotiating issues of practical importance. Zeno provides particular support to those responsible for moderating these kinds of discussions. The Zeno server is a Java application for the World Wide Web which enables and facilitates moderated issue-based discussion forums in a secure environment. Zeno discussion forums are integrated with a workspace facility for sharing classified documents).

Support to "Offline" Deliberation

The second project is Compendium which is based on IBIS to support face-face meetings. It facilitates substantive discussions, helps identify gaps and action items for follow-up, and leaves behind a record of the meeting so that the next meeting can build on it. Compendium can deal with issues where there is no objective problem definition but there are a lot of interpretations, there is no right answer but there are lots of tentative next steps. Compendium is based on the concept that collaboration among participants is the best approach to problem solving. One could imagine the situation where off-line consultation meetings could be facilitated with the support of tools like Compendium.

Mapping Online Deliberation

Work from the Social Computing Group at MIT Media Lab focuses on how to make sense of and visualise online conversations. One result from this research is *Conversation Map* which is basically a newsgroup browser designed to make it easier for participants to understand and reflect on very large-scale conversations like large, electronic-mail lists or busy Usenet newsgroups. The system analyses the content and the relationships between messages and then uses the results of the analysis to create a graphical interface. With the graphical interface, a participant can see the social and semantic relationships that have emerged over the course of the discussion. The Conversation Map system computes and then graphs out who is "talking" to whom, what they are "'talking" about, and the central terms and possible metaphors of the conversation.

Limits to Online Consultation

A question raised in any e-engagement initiative is the question of how representative the results of such online dialogue are. A major objection to e-engagement is that not enough citizens will become involved and that the technology is giving government self-selected comments. There is the risk that an active minority achieves an influence far beyond their number. Coleman and Gøtze argue that if a key objective of online engagement is to inform elected representatives then the selection of participants need not be concerned with representativeness and more concerned to engage a broad range of experience. However, given the current uptake of ICT, and specifically the Internet, in OECD member countries any e-engagement initiative should be seen as just one way to engage citizens. This typically means that there is a need to ensure multiple channels for citizen input and the need to ensure that e-contributions are fully integrated with offline contributions.

E-PARTICIPATION THROUGH E-PETITIONING.

A relationship based on partnership with government in which citizens actively engage in defining the process and content of policy-making. It acknowledges equal standing for citizens in setting the agenda, proposing policy options and shaping the policy dialogue – although the responsibility for the final decision or policy formulation rests with government

By allowing citizens only to decide between offered choices or comment on a fixed "menu" of options, government loses the opportunity for maximum input from citizens, while citizens are unable to express their true thoughts. Sclove concludes that:

> Decision-making processes are democratically inadequate, even spurious, unless they are combined with relatively equal and extensive opportunities for citizens, communities, and groups to help shape the decision-making agendas.

The OECD report *Citizens as Partners* states that few countries have started to address active participation, and that examples of good practice are rare. Therefore, although the emphasis in this section is on citizen engagement with national government we feel it would be useful to give a very positive example of how one particular parliament, the Scottish Parliament in the UK, is addressing e-participation through e-petitioning.

E-petitioning

In many countries around the world, citizens have long used petitions to make their feelings known about issues that concern them. Simply, a petition is a formal request to a higher authority, *e.g.* parliament or other authority, signed by one or a number of citizens. The format of petitions and the way petitions are submitted and subsequently considered by parliaments varies greatly (OECD, 2001, p. 19). Few countries have used technology to enable electronic petitioning to the extent of the Scottish Parliament.

One of the main documents setting out how the new Scottish Parliament should work was The Consultative Steering Group document. This stated that the Scottish Parliament should aspire to use all forms of information and communication technology "innovatively and appropriately" to support the Group's guiding principles of *openness*, *accessibility* and *participation*.

On the issue of petitions, the Consultative Steering Group stated:

> "It is important to enable groups and individuals to influence the Parliament's agenda. We looked at a number of models in other Parliaments for handling petitions and concluded that the best of these encouraged petitions; had clear and simple rules as to form and content; and specified clear expectations of how petitions would be handled."

To achieve this, the Scottish Parliament established a dedicated Public Petitions Committee (PPC) to actively promote petitions as a means by which the public could effectively raise issues of concern with the Parliament. The remit of the PPC is to consider and report on whether a public petition is admissible and what action is to be taken on the petition. There are no restrictions on who can submit a petition. A petition submitted by an individual will be considered on equal terms with one submitted with a large number of supporting signatures. The PPC considers the merits of the issues raised in each admissible petition and makes a decision on the appropriate action to be taken in each case. This can involve requesting other committees in the Parliament (generally those with the remit to examine specific subject areas) to carry out further consideration of the issues raised, or requesting the views of, or action by, the Scottish Executive (the devolved government for Scotland), local authorities and other public bodies in Scotland.

The electronic petitioning system used by citizens to submit petitions to the Scottish Parliament is called "e-petitioner". An in-depth description of e-petitioner is published elsewhere, and it is only necessary here to explain its main "participation" features. The e-petitioner tool is designed to:

- View a petition text online.
- Read additional information on the petition issue online.
- Those deciding to support the petition can add their name and address to the petition online.
- All citizens can join an integrated online discussion forum and add comments for or against each e-petition.

Precise guidelines produced by the PPC in relation to petitioning the Scottish Parliament provide citizens with help in creating appropriate e-petitions. Management procedures have been put in place to incorporate the submission of e-petitions into the normal workflow of the PPC. A briefing note accompanies each e-petition. This contains the petition text, a list of names and addresses of those supporting the petition. An analysis of the geographical spread of supporters is included so that MSPs can see how many of their constituents support the petition. Also there is a summary of the discussion forum highlighting the main arguments for and against the petition.

Finally, the Committee ensures that petitioners are kept informed of progress at each stage of the Parliament's consideration of their petition. This feedback is reflected on the e-petitioner website. The actions of the Committee have resulted in a range of positive outcomes, from local solutions to petitioners' concerns to amendments to legislation – thus demonstrating active participation.

Online Referenda

Referenda, initiated by citizens, is an established way to introduce items onto the political agenda and influence policy content in some OECD member countries. The development of online referenda is the natural extension to this. Depending on legislation in the specific countries, this may necessitate similar levels of security and citizen authentication that is needed for legally binding e-voting. Several voting technologies, *i.e.* by SMS, interactive digital TV, electronic kiosk voting and remote electronic voting are potentially applicable. However, the design of the system would have to accommodate the different extent to which sophisticated user interfaces could, or could not, be developed.

E-petitions and e-referenda are just two of the potential mechanisms for e-participation. Participation in designing public policies can also be influenced by analysis of, for example, emails, FAQ lists, chat rooms, discussion forums, and online "visiting hours" for elected representatives. The development of "online communities" of interest, in which specific policy issues are debated and alternative proposals formulated, are also promising examples of active participation online.

References

Becker, T. and C. Slaton (2000), *The Future of Teldemocracy,* London.

Blumler, J.G. and S. Coleman (2001), *Realising Democracy Online: A Civic Commons in Cyberspace,* London.

Castells, M. (1996), *The Rise of the Network Society, The Information Age: Economy, Society and Culture*, Vol. I, Oxford.

Rose, R. (2002), *The Internet's Impact on East Asian Governance: Openness, Impersonal Rules and Accountability*, Glasgow.

Zolo, D. (1992), *Democracy and Social Complexity: a realist approach,* Oxford.

7

Open Source Solution for E-Governance

Open-source software (software with open source code) has become very popular in the last few years and is advancing at a speed unknown outside the world of IT. Just a few years ago, open-source software was regarded as a slightly 'nerdy' rebellion against the giants of IT. Today it appears in television advertising in the company of the IT giants. In other words, open-source software has become an area of business – an alternative, and therefore a competitor, to proprietary software.

This interest has also spread to the world of politics. Because open-source software is now increasingly used for commercial purposes, because it is characterised by independence from software producers, because it is opposed to the creation of monopolies and because it is characterised by a 'free-of-charge principle', open-source software has had a great impact on the political agenda, nationally and internationally.

Open source designates software which is developed and maintained according to principles of far-reaching user rights, which include the right to alter, copy and distribute the software. This software policy is of interest today, because development of IT over the last ten years has been dominated by the Internet, and the Internet is fundamentally made up of open-source technologies. The expansion of the Internet has made this network the hub for the information exchange of both enterprises and authorities and increasingly also in important work processes.

Open source today comprises more products than those linked closely to the Internet. There are open-source software, operating systems, cooperation systems and special systems. At the same time, many of these systems benefit from the Internet, which makes possible digital cooperation between geographically separated units.

The public sector in Denmark and the whole of the EU faces a change-over to digital communication in the supply of public services, with the citizen at the centre. This means that coherent services have to be delivered to a greater extent, necessitating increased cooperation between administrative units and between the levels of the administration of the EU, national government, county and local authority.

This development makes great demands on the IT systems on which e-government is based, for instance with regard to options for exchange between authorities and with citizens and on the security with which this exchange takes place.

Open-source Software in E-government

A change-over to e-government will necessitate huge investments in IT over the next few years. It is therefore natural that a close assessment is made in connection with these investments of the forms of information technology it is intended will be applied, and who controls the development and ownership of the fundamental technologies in e-government. This increases interest in the opportunities opened up by open-source software, and makes the question of the potential for the use of open-source software in e-government economically advantageous and relevant.

E-government and the Requirements to be met by the Public Sector

The development of information technology over the last decade has left clear marks on society. In particular, the emergence and expansion of the Internet has revolutionised the way we communicate and affected organisation and interactions in society. The public sector has also 'gone on-line' – at first with general information on institutions and the work of authorities. But gradually, as the public and companies have become more accustomed to making bank transactions and shopping through the Internet, it has been increasingly expected that the public sector is also capable of supplying more complex public services. As a result of technological development in the public sector in general, it is said that 'e-government' is being developed. This can be described in practical terms as follows: 'Use by public administrations of technology, primarily Internet-based technology, which contributes to improving access to and supply of public information and public services to citizens, companies and the public sector as a whole'.

It is therefore expected that e-government will have the potential to create easier and smoother interaction between the public sector and the citizens, but are still at the beginning of this development. A model presenting a schematic overview of development in e-government is shown below:

The vast majority of public institutions now have their own websites, with general information on the aims and work of the institution, which relates to the first phase in the model: The existence stage. Many public institutions have gone a step further to the interaction stage and today additionally offer more advanced websites, where the content is adapted to a greater extent to the individual user, and where users are able to interact with the websites to some degree. The users (citizens) have the option, for example, of downloading forms, information, guides, can carry out information searches, communicate with institutions by e-mail, and so on. Some public authorities have reached the transaction stage and offer websites where actual transactions can be made. This may involve students, who have the option of enrolling for university courses on-line, or citizens submitting their tax returns through a tax authority website.

Public administrations for centuries have been made up of vertical units, which supplied a particular product or service to society. If a particular task was to be accomplished as a citizen or enterprise, it was necessary to apply to the series of authorities dealing with the task concerned.

As a result of technological development and citizens' increasing expectations for the level of public service, there are now calls for government service systems to be put together on the basis of public's need for a readily accessible and logically coherent chain of transactions. It has to be possible for services to be provided through one point of entry, an Internet portal, which integrates all the services which members of the public or enterprises might need in a particular situation. It may be a portal aimed at families with children, which offers all the services a young family requires, regardless of which public body is to supply the service.

Portals of this kind call for different connections in the procedures of the various public institutions, and only a few public services are supplied in this way at present. The *transformation stage* requires significant changes in the public sector, not just in the construction of IT system, but above all in procedures. e-government is therefore still some way from fully utilising the potential that exists in the use of information technology and the Internet in particular.

E-government in Denmark

Development is, however, well under way, and e-government comes high on the political agenda, including in Denmark. This was recently emphasised by the two party leaders, the Prime Minister, Anders Fogh Rasmussen (Liberal Party), and the Minister for Economic and Business Affairs, Bendt Bendtsen (Conservative People's Party): 'The Government wishes to make it easier for members of the public to deal with the authorities. This will be one of the areas focused on by the Government in the future [...] ... 'Members of the public must be able to contact public authorities 24 hours a day through the Internet. Businesses must be able to obtain all forms and information from one place [...]. Young families must also be able to find information on child allowances and childcare facilities in one place,[...]. Members of the public must be able to see their own data in public registers and be able to track their cases in the public sector on the Internet. This creates greater openness and more democracy.' In May 2002, the Government published its modernisation programme, which states:

'The Government wishes to put Denmark at the forefront as a modern IT and knowledge-based society and apply new technology to alter the way work is done in the public sector. The aim is to provide the public with a better service and release public servants for tasks which make a direct contribution to public welfare. Efficiency improvements and less bureaucracy are achieved when the entire accomplishment of the task is thought through. The Government therefore wishes to explore new avenues in all areas and challenge established procedures and paper-based routines... A commitment must be made over the next few years to digital solutions that do away with manual routine case processing and reduce administration. The outlook is for more efficient public procedures and faster and error-free processing of cases, to the benefit of the public and businesses....New technology must help create increased cooperation across the public sector and traditional sector boundaries. Taking account of the legal rights of members of the public, it must be ensured that information can be exchanged between public

IT systems, so that members of the public experience the public sector as being an efficient entity which works smoothly. In that way, duplicated work is avoided and members of the public do not need to provide the same information repeatedly.

From centralism to decentralism and back again

From the historical point of view, development in IT in national government has gone from centralisation to decentralisation and is now on the way back again. The administration of IT purchases by national government was originally the preserve of the Ministry of Finance, which had to approve all purchases. As a result of a rising number of purchases and investments in IT, it was not possible to approve all purchases, and in 1992 the approval procedures were therefore abandoned for national government IT purchases, and IT investments by national government were in reality made freely.

The report 'Info 2000' from 1994 recommended that national government should start discussing how IT could be developed in the state sector, and provided the basis for development of an actual IT strategy. In 1994, major tasks relating to IT in national government were transferred to the Ministry of Research from the Ministry of Finance, resulting in the preparation of annual action plans for IT development and application.

In the autumn of 2000, the Ministry of Finance appointed the 'E-government Committee', an interdisciplinary committee charged with ensuring the development of e-government across existing organisational boundaries, ensuring better and cheaper public service and improving the efficiency of procedures by introducing e-government. The Committee was composed of representatives from the National Association of Local Authorities in Denmark, the Association of County Councils, Copenhagen Local Authority, Frederiksberg Local Authority, the Ministry of IT and Research, the Ministry of Business Affairs, the Ministry of Internal Affairs and the Ministry of Finance, which also provided the chairman. Together they made up the Digital Task Force.

E-government according to the Digital Task Force

The Committee published a report on e-government in 2002. The vision for digitised public administration was defined as: 'improved and more efficient accomplishment of administrative tasks through the application of information technology to the benefit of members of the public, businesses and the public sector.'

In lending weight to IT-supported development of procedures and organisation, e-government differs from IT application in the public sector to date, which has primarily entailed technical support of existing procedures.

It is important to point out that e-government, which both definitions establish, is concerned with improving the level of service to society using technology. The report from the Digital Task Force has identified four pointers which should guide work on e-government. The four targets are:

- Target 1: e-government should actively contribute to the development of network society: The public sector can do this by virtue of its size, for instance by being involved in the setting of standards for IT use and increasing IT skills broadly among the population.
- Target 2: The public sector should work and communicate digitally: This involves digitising work processes in relation to members of the public and companies.
- Target 3: The services of the public sector should be delivered in a coherent way with citizens and companies at the centre: It should be possible for citizens to communicate with the public sector by accessing one place. It should be possible to access this place 24 hours a day, 365 days a year.
- Target 4: The tasks of the public sector should be carried out where they are best dealt with: e-government should break down administrative boundaries and ensure that specific work tasks are handled flexibly in a more dynamic organisation. Sharing of knowledge between the different parts of the public sector should be expanded using information technology. As work processes are digitised, not only will processes be changed, there will also be a change in the organisation in order to raise the level of service and efficiency.

E-government in practice

A large number of projects of varying size are already under way in the public sector to promote the development of e-government. Some examples are:

Project	*Contents*
Digital Signature	The aim of the project, in conjunction with private players, is to establish the necessary and sufficient technical and organisational frameworks for the introduction and use of digital signature on a broad basis in society.

Family at the Centre	This project is to analyse the present situation and propose coherent digital solutions for services and situations, where young families (children 0-6 years) are in contact with the public authority, for example in connection with childbirth, day nurseries, maternity leave, childcare, child benefit/allowance and nursery school, and contact between home and institutions.
E-Citizen	In cooperation with the National Association of Local Authorities in Denmark and the Danish State Information Service, this project is to assess the need and opportunities to create one or more attractive interdisciplinary common citizen portals. These portals will contribute to better citizen service and improved efficiency in the public sector.
Collective integration process	Increased cooperation with smooth exchange of information between local authorities and the central immigrant integration authorities will improve both quality and the efficiency in the individual immigrant's contact with the public sector.
From Hospital to Home Care	This project will improve electronic communication between the local authorities and hospitals in connection with admission and discharge of patients, so that a more holistically oriented effort can be made.
Industrial injury cases	Contact takes place with many parties in the processing of industrial injury cases. There is great potential in digitising this interaction. There is much to be gained in particular in the interaction with the municipalities by re-using medical assessments which today are requested by both local authorities and the National Board of Industrial Injuries. The aim is for the citizen to have to supply information only once.
One account for enterprises in the public sector	This project is intended to prepare for a decision on collective account for public-law payments by enterprises and financial holdings with the whole of the public sector. This will mean large efficiency gains for both the public sector and enterprises – particularly if terms of payment, interest rates etc. are harmonised at the same time.

Information Technology Requirements of E-government

Such extensive dependence on information technology makes it necessary to consider carefully which technologies are selected for use in e-government. What are the total expenses, both direct and indirect, in using a particular technology? How does a given technology work together with other technologies that are used in connection with e-government? What opportunities and rights does the user of the technology have? How well does a technology work? How is data exchanged most smoothly between the systems? What are the social consequences of a given choice of technology?

Security is the key

Security is one of the key topics in relation to e-government. Increased use of technology means that many items of information will be stored electronically, and it is therefore necessary to rely on the software used not containing elements that permit external access or opportunities for undesirable release of information to third parties.

Finance and efficiency

The aim of e-government is to improve the service provided to society, for example in the form of more efficient electronic processing of cases. Improvements in service via the Internet at the same time mean financial savings for the public sector in the longer term. Some key topics in relation to finance and the debate on efficiency are presented below.

Work processes

The aim in introducing full e-government, as mentioned, is not just to add a technological solution to existing public administrative routines and procedures. Efficiency gains in introducing e-government are dependent on work processes being re-assessed and modified. Information technology therefore has to support adjustment and simplification of existing work routines. Against the background of a survey of 65 government institutions, the Ministry of Finance has calculated that there is efficiency improvement potential of between 17 and 31 per cent in administrative resources in changing over to electronic administration. This is equivalent to between 2 and 17 per cent of total operation, depending on the type of institution concerned.

Integration of systems

But the potential for efficiency improvements depends just as much on the extent to which the internal administration in government is integrated with the systems that are in contact with the world around and this integration proceeding across existing organisational boundaries. If e-government is to be effective, it is necessary for the systems to be able to communicate with one another. There is a need to standardise such solutions so that it is possible for systems to be integrated and data re-used. As we shall see later, open standards are essential for such development to succeed. It is therefore appropriate to focus on how public data is exchanged and stored.

Management

Several public institutions will shortly need to develop new IT systems. A key question in relation to both finance and efficiency is how these systems are to be developed. Broadly, two questions can be asked on the management of development: is it to be done using central management, which may mean that the system is not adapted to local needs, or is the development to take place in a decentralised way, which may mean that the same work is performed and paid for several times?

National government as purchaser

A close relationship may arise between supplier and buyer in the purchasing of software. This often leads to dependence on the supplier, often referred to as a lock-in effect. Once a system has been acquired, the user is dependent on when and how the supplier modifies the product. A number of supply rules have been developed to counteract lock-in, but the question of whether these are adequate remains. Consideration therefore has to be given to how national government, as the purchaser, is to act in relation to possible suppliers of systems, including questions on updates and enhancement of the software.

National government as a financial player

The size of the public sector in western European societies means that public provisions are of great significance to the private business community. One area in which this applies is procurement of technology. In making purchases, authorities are subject to a requirement not to favour individual suppliers unfairly. At the same time, the size of public purchases, if they are coordinated or merely coincide, means that selected products may gain market-leading status in the Danish market depending on the extent to which the product contains imports or exports. The investments necessitated by the concept of e-government could therefore have effects on market structures and the development of prices on Danish markets, while they will not significantly affect global market products.

Open-source Software in the Public Sector

To date, the public sector has mainly made use of proprietary software. One of the many technology choices the public sector now faces is the question of whether open-source software has sufficient functionality and user-friendliness in comparison with proprietary software and whether it is cost-effective from an overall perspective.

Open-source software is not necessarily free, but in most cases is considerably cheaper than proprietary software. There may therefore be economic advantages for the public sector in using open-source software.

Open standards and possibility of integration

The characteristic feature of open-source software is that it uses open standards. The expression open standards means that the principle for the development of software is established in public forums, in contrast to proprietary (industry) standards, which are kept secret. It also means that it is often possible, through a more or less democratic mechanism, to influence the standard. With e-government, it is necessary to be able to read and exchange data without encountering problems. It is therefore appropriate to focus on how public data is exchanged and stored.

Possibility of access

Many items of information in e-government will be stored electronically. It is therefore necessary to rely on systems not containing elements which permit external checking, or undesirable release of information to third parties. Systems with freely available source enable both government and other experts and members of the public to carry out post-inspection of the programs used. It is therefore not possible, for example, to hide what are known as 'back-doors', i.e. secret entrances, in an open-source software product, as these will be discovered when the code is inspected.

Security and quality

Open source code enables the user and other interested parties to check whether the program is written in a justifiable way and where appropriate to identify elements of danger for the stability and security of the code. This makes it possible for open-source software to be secure and of high quality.

Supplier independence

A 'lock-in' situation can easily arise when software is purchased: once one system has been acquired, one is tied to it and becomes dependent on when and how the supplier modifies the product. This 'lock-in' will not arise if open-source software, which is based on open standards, is used. Data is not stored in a proprietary format, and it is possible for users to change between several different systems and therefore also several different suppliers.

Possibility of customisation and re-use

As the source code is open and it is permissible to modify this, public authorities can adapt open-source software to particular needs. It will be possible for a customisation to be re-used in other parts of the public sector, resulting in lower software customisation costs.

Open-source Software from the Political Point of View

Open-source software has attracted attention in the political system. In 2000, the Research Committee of the Folketing, the Danish Parliament, adopted a report on 'Proposals for a Folketing resolution on a strategy for the expansion of open-source software in Denmark'. In this report, the Committee recommended that the Government should contribute through national IT policy towards providing information on the potential offered by open-source software and should recommend that the use of open-source software be included as an option in invitations to tender relating to IT. Several European countries have launched initiatives on the use of this type of software. This has meant either replacing existing systems with open-source software or making recommendations on future use of open-source software.

Germany

The German central administration in June 2002 entered into a framework agreement with IBM and the firm of SuSe on the supply of open-source products, based on Linux. The agreement makes it possible for German public administration to acquire Linux-based systems at a reduced price from IBM. The agreement makes it possible to supply servers, as well as Linux-based workstations. IBM will provide ongoing support for use of the systems. The German Government wishes to promote alternatives to Microsoft with the contract, but this is not a law, more an offer to the public decision-makers.

UK

The British Office of E-envoy issued a British policy in the area of open source at the end of July 2002. It is stated in the open-source policy that the British Government and British authorities will in future consider open-source software systems on an equal footing with proprietary solutions when purchasing IT. It is also an objective that the British Government will in future as far as possible use products based on open standards. In principle, the British Government wants to get 'as much as possible for its money' in IT purchases, which is one of the arguments for considering open-source software. In addition, the UK Government wants to avoid problems in the future with 'lock-ins' in relation to a specific supplier.

France

The French Government has decided that French central administration should terminate its agreement with Microsoft on the supply and use of their software. This decision means that all French national and local authorities as far as possible are to use open-source software. This is software where the program kernel is to be publicly available, and where the finished programs are freely available and at no charge, for instance via the Internet.

An office, the Agency for Information and Communication Technologies in Administration (ATICA) was set up in August 2001 to follow up the French Government's decision, to coordinate the IT initiatives in and between the public authorities. The Agency is to ensure that public IT projects make use of open source standards with a view to ensuring interoperability and reducing expenditure on IT.

Finally the French Government wishes to improve the opportunities for small companies in the software area, by making it possible for them to work on public open-source projects. The aim is to support the enhancement of open-source software.

Other Countries

Several open-source software initiatives have seen the light of day both in Europe and in the rest of the world. In Finland, some members of parliament led by Kyösti Karjula have issued a recommendation on the use of Linux platforms in public administration. A working group appointed by the European Commission has issued a report that recommends public administrations in the Member States to use and swap experience relating to open-source software. Finally there has been some discussion in Peru on a legislative proposal compelling public authorities only to buy and use open-source products. The arguments in favour of the use of open-source software here are based on the possibility of access, the permanency of data and the security aspect.

What is Open-source Software?

Who would buy a tin of tomatoes with no product label on it, or who would buy a car with the bonnet welded shut? These are the kind of rhetorical questions advocates of open-source software ask to emphasise that the purchaser of proprietary software has too few rights. When software is open source, the source text is available, the user has the right to modify it and several users can exchange improved versions among themselves. In traditional software, which we refer here as 'proprietary software', users are unable to modify the software themselves. When there is a need for bugs to be fixed, for example serious security defects, users are dependent on the ability and willingness of an individual supplier to supply these -like the unfortunate owner of a car, where the engine can only be repaired in the supplier's workshop and on the supplier's terms.

Open-source software is thus linked to the question of rights to software. This chapter introduces open-source software by first outlining how it relates to a practice determined by attitude and movement among developers and users of software. The remainder of the chapter describes rights as laid down in licences for open-source software. The licence for the open source web server Apache is examined by way of example. There then follows a general overview of the rights the user has to open-source software. Acquisition of open-source software is generally – but not always – free of charge, and this is looked at in depth, after which there is a discussion of the relationship between open source and open standards. The statutory basis for software licensing consists of copyright legislation, which is finally outlined, together with an explanation of the working group's choice of the term 'proprietary'.

Open Source as Practice and Movement

The Free Software Foundation is the best known of the early organisations that worked to promote what is now referred to as open source. The Free Software Foundation (FSF) was set up in 1985 by Richard Stallman to support the development of a range of software including operating system, compiler and editor. The aim was to create such an extensive software package that users had a complete productive desktop at their disposal, without any proprietary software.

The Open Source Initiative (OSI) was formed in 1998 to establish a pragmatic alternative to FSF. The people behind OSI, including Eric Raymond, regarded FSF as ideological and confrontational, and wanted to establish a platform to disseminate the open source ideas in a more pragmatic way, particularly towards the IT industry.

Sharing of Software and Knowledge

A common attitude in the open source community is that the restrictions on the rights of users of proprietary software prevent the sharing of software and knowledge, and that this conflicts with the user's interest but also in a broader sense with the objective of promoting the development of software technology.

Another important element in attitudes in the open source world is support for open standards. The HTML standard is an example of an open standard. A particular characteristic of open standards is that their definitions are publicly available in contrast, for example, to Microsoft's doc file format, which is secret. Open standards promote user independence. For example, anyone can attempt, on the basis of the definition of the HTML standard, to develop programs that write or read HTML documents (e.g. web browsers).

The Open Philosophy of the Internet

The development of the Internet from its beginning in the 1980s and 1990s was in several ways linked to the open-source philosophy. The central network protocols etc. were developed in an open standardisation process under the auspices of the World Wide Web Consortium and the Internet Engineering Taskforce, partly on the basis of experience gathered from reference implementations of the standards. These were programs in which the source text was available and were distributed under open-source-like licences.

Part of the development work had roots in universities and their tradition of openness and publishing results. The Internet, with electronic mail and newsgroups, was the communication channel for the groups who in the nineties developed open-source programs such as Apache and the operating system Linux. The successful development of these programs showed that extremely complex programs could be developed as open source. A generation of software developers, network people, system administrators etc. acquired positive experience of the spread of technology based on open standards and available source text by participating in the development of the Internet.

Patents

Advocates of open source have entered the debate on the introduction of patents on software, which is regarded as a disincentive to the development and spread of new knowledge in the area of software. Software patents can create difficulties for open-source projects, as these projects rarely have financial funds at their disposal to pay for patented methods.

Open Source can be Commercialised

Is open source compatible with commercial business strategies – or does the concept stand and fall with idealism and volunteer labour? As an open-source licence entitles the user to copy and distribute the software, the supplier's ability to sell licences is undermined. By far the greater part of the costs of software is accounted for not by the original development of the software, however, but by services in the form of customisation and maintenance, and the open source model is not an obstacle to firms selling such services.

Many private companies are involved in the development of open-source software. Sun and Netscape have implemented large open-source development projects (the office package OpenOffice.org and the browser Mozilla respectively). IBM contributes to the development of the open-source web server Apache and the operating system Linux, and in 2000 announced that the firm planned to spend 1 billion US dollars on developing open-source software.

How far has Open Source Come?

The development of open-source software has reached the stage that today it is possible to create a fully functional workstation that is based solely on open-source software, and that can be used as a tool in advanced software development – or for word processing and other functions used by ordinary users in the public sector. It is also possible to build up all the functions included in a modern website. The purpose is to analyse how the public sector can make the best possible use of the potential in open-source software with different forms of licence, and in each case choose the software that provides the best solution for a particular task.

The User's Rights According to Open-source Licences

Open-source licences give the user:

- access to the source text
- the right to use
- the right to modify
- the right to distribute

In the view of the working group, this definition captures the essence of what is generally understood by open-source licences. The definition is consistent with, but less extensive than, the one given by the Open Source Initiative (OSI).

The organisation runs a certification service on the basis of its definition, and has approved the Apache licence and more than 40 others. One of the functions of OSI is to shed light on those licences that have been used when firms have released software already developed as open source, and it has sought to establish open-source projects led by the firms themselves. In this situation, OSI's assessment of the licence may be of interest to potential participants outside the firms, who have wanted to make sure that the often complex licences were 'proper' open-source licences.

The Free Software Foundation early on developed what is known as the GNU licence, which is more far-reaching than the minimum requirements in the OSI definition. The GNU licence requires that modified versions of the original software are also made available to other users under open-source licence terms. In contrast, licences such as the Apache licence allow for a firm to 'close' an improved version of Apache and try to sell traditional licences. These differences are substantial for firms that work commercially with open source, but not for government as a user of software products.

Access to the source text means that the software can be 'reviewed' by experts who are independent of the supplier. When software is only supplied in binary, machine-executable form, on the other hand, it is not possible in practice to recreate the source text, and it may also be prohibited under the software licence. Information on the features of the software can therefore for the most part be obtained only by studying how it behaves during execution, which gives far less of an insight than if one also has access to the source text.

The right of use means that no obstacles are put in the way of using the software for example for commercial, military or other purposes. The right of use also means that open-source programs may be used together with traditional, proprietary software. A public organisation is therefore at liberty to choose a 'hybrid' strategy, where both open-source and proprietary software are used.

The right to modify gives the user organisation the right to carry out any form of customisation, debugging etc. of the software it might want. The right to modify is linked to access to the source text, without which it would not be possible in practice to make a desired modification. The right to modify can be utilised internally by IT staff, but also makes it possible for suppliers to be selected on the basis of such service tasks. The right to modify consequently provides supplier independence in relation to maintenance tasks.

The right to distribute means that the user is permitted to create several copies of the software and to distribute them, both internally in the organisation and to other organisations.

Distribution and Payment

A consequence of the right to distribute is that open-source software can be acquired at distribution costs, which means either free of charge or for a small charge, for example for dispatch of a CD. Open-source licence rules do not, however, contain any provisions relating to the software having to be free of charge. The Apache licence, for example, does not contain any prohibition on accepting money for the Apache software, or for products containing Apache. The question of payment for open-source software is covered by the first item in the OSI definition, which requires that open-source licences do not prevent or impose such payment: 'The license shall not restrict any party from selling or giving away the software as a component of an aggregate software distribution containing programs from several different sources. The license shall not require a royalty or other fee for such sale'.

When open-source software typically is free of charge, this is consequently merely a mechanism based on the right to distribute. This mechanism presupposes that the software exists beforehand, and is distributed to a customer who is willing to distribute it further. If the software does not exist in advance, on the other hand, and the customer therefore places an order with a supplier to develop it, the customer obviously has to pay for the development work, whatever the form of open-source licence.

On the basis of the compatibility of open-source licence rules with accepting money for the distribution of open-source software, there are firms that have this as part of their business concept. The firm Red Hat, for example, charges money to send a pack containing a number of CDs with the open-source operating system Linux and associated extra software.

Under the software licence, the purchaser of one of these packs can both freely copy the software within his organisation and make it available to others as copies on CD or via the Internet. The firm Red Hat and other distributors of open-source software will therefore only be able to induce users to pay a price equivalent to the convenience of having a ready-made CD instead of having to burn one, the certainty of knowing that the CD actually contains the particular software suite Red Hat has selected and other accompanying products, such as a book containing documentation on the software.

In addition, the purchaser can choose to pay a premium purely out of sympathy for the open-source idea and to assist in securing an economic basis for the continued existence of the distributor, but this hardly constitutes a major source of income.

Open Standards

Open-source software preferentially uses open standards. Just as no principle of being free of charge is laid down in the open-source licences, neither is the use of open standards stipulated in them. The licences only regulate the relationship between the copyright-holder and the user, and do not describe in detail what type of standards or other methods are used in the software.

The standards concerned are definitions of file formats, layout, protocols etc. The standards typically define interfaces between the software and its surroundings, for instance standards for how the software transfers data to and from files on the computer, or in exchange with the programs of other computers via the Internet.

An open standard is understood fundamentally as being a published definition. To take some examples, HTML and TCP are open standards, because they are published - by the World Wide Web Consortium (W3C) and the Internet Engineering Task Force (IETF) respectively, which have been responsible for developing the standards.

In a broader sense, the term open standard contains some other requirements, including that a standard is developed in a consensus process, which is not dominated by a particular supplier, and that the standard is not merely public in principle but is available free of charge or cheaply. It is supplemented by the standards developed in W3C and IETF. Unless otherwise stated, however, we use the term 'open standard' here in the narrower meaning of a published definition.

It is logical that open-source software preferentially uses open standards, firstly because open standards are in tune with the basis of values in open-source philosophy, and secondly because the use of a standard in an open-source program in itself signifies partial publication of the standard.

There is, however, no clear connection between the licence form of the software and the standards the software uses. Firstly, proprietary software can use open standards. Secondly, some source programs actually use closed formats. OpenOffice/StarOffice, for example, uses Microsoft's doc format. This is normally done as a matter of necessity, to ensure that the software can be used together with existing, closed formats, and in the case of StarOffice/Open Office the product has its own, preferred file format, which is open.

Licences and Copyright

Copyright legislation is the basis on which the software supplier is able to use licences to stipulate rights (and obligations) for the user's application of the software.

Under copyright legislation, both in Denmark and in other countries, software is counted among works within the meaning of copyright law. Section 1 (3) of the Danish Copyright Act reads: 'Works in the form of computer programs shall be counted as literary works'. It is not normally the case that the user signs the licence agreement. The normal situation is that the user implicitly accepts the licence terms by starting to use the software. An installation program often draws the attention of the user to the terms of licence – and to the fact that putting it into use is deemed to signify acceptance of the terms.

Copyright to proprietary software normally belongs to a commercial enterprise, and Microsoft, for example, holds the copyright to the Windows operating systems. In the case of open-source software, the copyright-holder may, for example, be the Apache Foundation, which holds the copyright to Apache, as a result of the developers of the software having assigned their copyright to it.

We have chosen to use the designation 'proprietary software' here for software with traditional licences. None of the designations known to us appear to be entirely pertinent, and the contrasting of proprietary and open source may be misleading, as the firm or the developers hold the copyright in both cases.

More specifically, we understand by proprietary software that the licence

- does not provide access to the source text
- only provides very limited access to distribution, typically for purposes linked to the individual user's general use of the program, such as the creation of a backup copy

In the case of proprietary software, the licence may directly prohibit the user trying to recreate the source code on the basis of the binary form of the software, by decompiling or 'reverse engineering'. The Copyright Act lays down a limit on the extent of this form of prohibition, however, as far as interoperability is concerned. Section 37 stipulates that the user shall not be prevented from undertaking decompiling with a view to attaining interoperability. In other words, the user may try to recreate the source text for parts of the software, provided this is necessary to establish an interface between the program and other software.

Licences for proprietary software normally contain exclusions of liability fully equivalent to those contained in the Apache licence. A user who is 'hacked' and incurs large expenses as a consequence of a security loophole in the software cannot obtain compensation from the supplier.

There may be reason for the user of proprietary software to examine the more detailed provisions of the licence:

Microsoft licences for organisations with many users – known as volume licences – have a common part under the name of Product Use

Rights. The product use rights contain some controversial sections, which give Microsoft certain far-reaching rights in relation to the user:

- The rules entitle Microsoft to collect and store technical information about the user. The information may also be passed on to third parties, provided the user cannot be identified
- In the section on Internet-based service components, the licence stipulates that for some software components Microsoft has the right to automatically download upgrades and fixes to the user's machine

Microsoft's gathering of information on the user can be done via the Internet, as the installation, activation or use of Microsoft's operating systems or Internet browsers may mean that the software automatically establishes a connection to Microsoft. Skilled users can, however, presumably protect themselves against at least some of the measures the licence rules entitle Microsoft to take, for example by deactivating particular features, or ultimately deleting certain programs.

Although proprietary software today is the traditional form of software, the severe restrictions on the user's rights are a relatively new phenomenon. In the 1940s and 1950s, software was predominantly developed in research environments. Commercial software development took place particularly at hardware producers such as IBM, who developed software to be supplied together with the hardware, and the customer did not pay separately for the software. Software to be supplied in return for payment was rare, and was normally special software, developed on the basis of contracts with a single large customer.

It was not until the late 1960s that the business model for proprietary software arose in the United States. An industry of software suppliers had developed there who were independent of the hardware producers, and who produced for a market where there could be several customers who were interested in the same software. It was in this connection that the supplier's strong interest in having sole right to distribute and therefore sell software licences arose, and adaptation of copyright legislation etc. to cover the area of software followed from this. In the early phase of the history of computers, on the other hand, it was general practice for the user to have rights to the software along the lines of those given by open-source licences, which are therefore in a sense the true 'traditional' rights to software.

Categories of Open Source software

In analysing the economic consequences of increased use of open-source software in government, the working group will divide software into three categories:

- software on the desktop
- infrastructure software
- custom built software

Software on the Desktop

Desktop software is the software that administrative staff use in their day-to-day work. We are thinking here in particular of office software such as word processing, spreadsheets and other programs in office suites, as well as web browsers, e-mail clients and calendar programs. This category comprises both software in the form of independent applications, for example for word processing, and software in the form of clients, such as e-mail clients, which require connection to a server.

Examples of infrastructure software are web servers, mail servers and operating systems. The infrastructure software usually constitutes the underlying element on which e-government and modern network society as a whole are based. It is an electronic counterpart to the transport infrastructure in the form of motorways and railway lines. The most important part of the infrastructure is the Internet and the World Wide Web, which among other things connect citizens with administration.

We understand custom built software to mean software that is developed for government with a particular application in mind.

Software on the Workstation

Desktop software in the public sector is dominated by Microsoft products, including the Office suite with the word processing program Word. The dominance of Microsoft products means that Microsoft's formats represent the de facto standard for the exchange of word-processed documents in government. The most important of these formats is the doc format, used by Word. The doc format is referred to below as a file format. The format is used to store a document as a file.

In the short term, if open-source alternatives are to be more widely used, they need to be able to handle Microsoft formats. It is difficult to achieve this fully, because Microsoft formats are secret. There is therefore no free competition on workstations. This makes it desirable that open standards for the desktop should be introduced in the long term, particularly a file format for word-processed documents. An important question to be answered is what requirements should be set for a new standard in addition to being open.

The Apache developer Brian Behlendorf mentions a number of historical and cultural reasons why open source has been relatively weakly represented in desktop software.Behlendorf points out that open-source developers in particular have worked on developing the form of software they used themselves, and that this did not include word processing suites and other office programs. In recent years, however, a number of good open-source programs for word processing etc. have emerged.

Open Source Alternatives

It is the impression of the working group that StarOffice and OpenOffice.org are the two open source alternatives in office software that are of greatest interest to government. The two products are largely identical. OpenOffice.org is a purely open-source product. StarOffice contains OpenOffice.org as well as a smaller, proprietary part, including database software. Licences for StarOffice are sold by the firm Sun for a few hundred Danish kroner. We have only concerned ourselves in the working group with the (predominant) part of the StarOffice which is also contained in OpenOffice.org, and we consider it reasonable to regard the two twin products as one – 'StarOffice/OpenOffice'. StarOffice/OpenOffice is run as a development project by Sun, and contributions come from both Sun and independent developers.

StarOffice/OpenOffice is interesting because it is generally recognised that the product is of high quality, partly on the basis of independent tests and reviews. In addition, StarOffice/OpenOffice in our view is of particular interest because it:

- is a complete office suite equivalent to the programs in Microsoft Office, comprising among other things the word processing suite Writer, the presentation program Impress and the spreadsheet Calc
- uses XML as a basis for the product's own, preferred file format
- has good compatibility with Microsoft formats

Test

The working group has undertaken a compatibility test of StarOffice/OpenOffice, i.e. of the ability of the product to exchange documents in doc format and other Microsoft formats. The test involved the product variant StarOffice. The conclusion drawn from this test is that information is not normally lost on conversion (where loss means that part of the text disappears or changes), but that in some cases there is a loss of layout, particularly with regard to the location of layout elements.

It is difficult to decide whether the degree of difficulty of the documents in the text (e.g. the complexity of the layout) is representative of the documents exchanged between organisations in the government sector. It will be possible for most documents to be exchanged without problems, but in a minority of cases layout will be lost. A cautious estimate is that problems of some degree of seriousness will arise – most of which will be easy to overcome in 10-20% of documents. The compatibility problems which may occur in connection with decisions in individual organisations to switch to StarOffice/OpenOffice – while the rest of the government sector continues to use Microsoft formats – will therefore be surmountable, but will cause irritation and expense in the form of time spent on solving or circumventing problems with loss of layout.

It is difficult to imagine there being any open-source or other alternatives in the future that will be able to handle Microsoft formats completely free of errors. The doc format is a Microsoft business secret, and regular changes are made to it. Work on alternatives to achieve compatibility is therefore a constant uphill battle.

Overall, it can therefore be said that the public sector on the one hand benefits from there being a totally dominant standard. It is essential that documents can be exchanged without problems. On the other hand, competition between alternative office suites has to some extent been eliminated, because products other than those from Microsoft will inevitably face compatibility problems. This can therefore be described as a lock-in situation, which means that the dominance of Microsoft's products in the use of office software in government is maintained.

Open standards for electronic communication with the citizens

The question of open and closed formats on the desktop has another dimension with regard to communication between government and the public. As a matter of principle, it is unacceptable for citizens to be forced to choose office software, browser or other software from a particular supplier in order to be able to receive information from the public authorities.

According to rules laid down by the Danish National IT and Telecom Agency, government publications on the Net must conform to the HTML standard. There is no direct reference to the principle of open standards in the description of the objective of the rules. A more pragmatic concept of 'technical broad availability' is used instead.Ministries, agencies etc. today publish a large quantity of information, which is readily available to the public. At the same time, it is difficult to comply with the policy of open standards in external dealings with the public.

- *Finished documents*: the HTML documents published on government sector websites only fully comply with the HTML standard in a minority of cases.The reason is often that the tools used to generate HTML codes automatically do not do so correctly, and that correcting the errors manually is time-consuming. In many cases the relationship is of no practical significance, but in some cases it creates difficulties for citizens who depend on the provisions made for the visually impaired in the standard
- *Non-finished documents:* it is common for the public to be obliged to have software that can handle Microsoft formats if they want to read more provisional documents, for example in connection with informal consultation procedures, or if they want to undertake electronic exchange of documents, for example with tax authorities.
- Danish State Radio's service providing direct transmission of radio broadcasts over the Internet cannot be used by listeners who use the operating system Linux on their PCs, for example. DR has chosen Windows Media Player because it is free to the institution. There is an alternative product, 'RealPlayer', which can be used on Windows, Mac and Linux, but DR has to pay to use it. DR has been criticised by the open-source association SSLUG, and justifies its decision by pointing out that only 0.7% of DR's listeners use Linux.

Opportunities and limitations in an open, XML-based standard for word-processed documents

The current difficulties in fulfilling the objective of open standards are serious in themselves. They are also an indication that with continued digitisation – which requires integration between a large number of systems and therefore a decision on which interfaces are to be used between the systems – the question of the use of Microsoft formats in particular will become increasingly pressing, both internally in government and in relation to the citizens.

The key area to try to influence with a strategy for the use of open formats on the workstation appears to be the file format government uses to exchange word-processed documents – at present the doc format from Microsoft Word.

Firstly, the format for word-processed documents is more central than the formats for the other facilities in office suites (spreadsheets, presentation programs etc.), simply because it is predominantly documents of this type that are exchanged. Secondly, it will be far easier for government to ensure that the external electronic exchange with the citizens takes place on the basis of open formats if government itself uses open formats in its internal exchange.

It is generally recognised as a matter of principle that e-government should be based on open standards. This is emphasised, for example, in the report 'Digital Forvaltning'. The advantage of replacing the doc format with an open format in particular is that it will create real choice, with the consequence of greater competition on price and quality, and better conditions for choosing open-source alternatives.

Another condition to be met is that government can continue to undertake internal exchange of documents without major compatibility problems. It is therefore important to try to define strategies that provide both benefits, i.e. supplier independence and a format that can be used by all public organisations.

In principle there is nothing to prevent standards for the exchange of office documents being open. The development of the Internet shows that open standards for extremely complex technical circumstances may be the basis for communication between programs supplied by independent producers (for example of web and mail servers and browsers). The present situation in the desktop area of 'standardisation by monopoly' is therefore not the only way of achieving standardisation.

The working group first enumerates some basic requirements which a format for word-processed documents has to fulfil. These requirements mean that existing, widespread formats such as HTML and PDF are not suitable. We then discuss the options for basing a standard on XML. A standard for a format for word-processed documents has to fulfil the following requirements:

- The standard has to cover the layout of the document. It is not sufficient to be able to exchange texts without layout, for example as normally happens with ordinary electronic mail. Plain text (ASCII text) by definition does not fulfil the requirement of being able to reflect layout.
- The layout of the document must be independent of settings, type etc. of the software used to display the layout on a screen or print out the document. For example, a Ministry's logo must remain at the same place in the document. The HTML standard is therefore unsuitable, because the layout of an HTML document depends on the type and setting of the individual HTML browser, e.g. the size of the browser window
- The recipient of a document has to be able to modify the document. It therefore has to be possible to undertake electronic exchange of non-finished versions when cooperating on the formulation of a document. The format must therefore not be limited to being a final format. The PDF format is consequently not suitable.

The above requirements are far from exhaustive. A collective specification of requirements will be very extensive, as will the work on developing it, because it is not stipulated what it should cover. Among the other requirements a standard has to meet, word processing programs have to be able to save a document quickly. The above requirements are emphasised because they show that the existing, widespread formats are not sufficient.

XML standard

On the other hand, there is much to suggest that XML (Extensible Mark-up Language) is well suited as a starting point for a standard for office documents. XML and associated standards are being developed by the World Wide Web Consortium. XML is the basic standard, and the consortium is also working on the development of a range of associated standards, among other things to define transformations between different XML-based formats and to define layout.

XML is generally recognised from the outset as a valuable starting point for standardisation. In the report 'Digital Forvaltning', XML is proposed as a key element in future standards for the exchange of data in government. XML has additionally been chosen as the starting point for some public standardisation projects, for instance in the healthcare sector and in state administration.XML and associated technologies are attractive, because they are based on open standards and because XML has already gained international momentum through the adoption of the principles of XML and the incipient application of the technology. It can therefore be expected that there will be a very large supply of software that supports XML.

It must be borne in mind, however, in the discussion on using XML as a starting point for a standard for word-processed documents, that it is a long way from the present agreement on the advantages of XML, purely in principle, to the point at which government actually has a usable standard. There are several reasons for this:

- XML and associated standards developed by the World Wide Web Consortium cannot in themselves constitute a standard for a suitable format. A standard of this kind can, on the other hand, be defined with the aid of XML and associated technologies. No such definition work is taking place under the auspices of the Consortium, and it also appears to be outside the Consortium's remit to work on such a specific application of XML
- The XML-related public projects under way in Denmark do not cover the development of an XML-based standard for word-processed documents. The projects referred to earlier in this section, on the other hand, deal with documents with some fixed structures (e.g. patient records and various forms), where an advanced layout is unimportant or of secondary importance
- Although XML and associated technologies are open, it is possible to embed an existing closed standard in XML without changing the closed nature of the original standard. For example, it is simple to define any document – including the fact that it is in a closed, binary form – so that it becomes what is referred to as a well-formed XML document (this can broadly be done by adding what is known as a 'tag' at the start and end of the document)
- Finally, XML in principle is geared towards the structure of data, not towards layout. The orientation of XML towards data relates to the fundamental idea behind XML of separating structure and presentation,

in which XML differs from HTML. The purpose of the Word Wide Web Consortium's standard XLS-FO (Extensible Style-sheet Language Formatted Objects) is to supplement XML with a method to link a particular layout to a document, but it is still in a relatively early phase of development

The consequence of this is that before the government sector can introduce an XML-based standard for the exchange of word-processed documents, it is necessary either to choose an existing standard (which is therefore not 'official' in the sense developed by the World Wide Web consortium), or to enter into or implement a project to develop a standard. It does not appear reasonable to tie oneself to an XML-based format exclusively having to make use of the range of standards developed by the consortium, but to the extent that other standards are included it is very important that these are open, so that the complete document standard is open.

A candidate for an XML-based standard for word-processed documents is the one used by StarOffice/OpenOffice. The standard is open in the basic sense, where the definition is publicly available, but also in the broader sense, in thatthere is a reference implementation of the standard with available source text (namely StarOffice/OpenOffice itself).

The working group cannot assess whether the StarOffice/OpenOffice format is appropriate (as well as being open itself), but some fundamental conditions apply in the format, which are general for an open, XML-based format developed in connection with a particular office suite.

On the one hand, as a result of the openness of the StarOffice/OpenOffice format there is a good starting point for competing suppliers to develop programs that convert to and from the format (known as filters). For example, Microsoft can develop a filter between the StarOffice/OpenOffice format and Microsoft's own closed formats, without having to use any form of 'reverse engineering'. On top of that, due to the widespread use of the XML standard there are some tools that facilitate the development of such conversion programs and, in the future, of programs that convert office documents stored in the StarOffice/OpenOffice format to future formats. Use of the StarOffice/OpenOffice format therefore does not introduce a lock-in and dependence on closed formats in the same way as when Microsoft's office programs are chosen.

On the other hand, it is true that if the format is established as a standard with the same dominance as the Microsoft format has today, StarOffice/

OpenOffice will hold a preferential position in relation to competing products.

This preferential position (and therefore weakening of the competition between different suppliers) is due firstly to the fact that the format may be assumed to be adapted to particular features of StarOffice/OpenOffice. Secondly, the continued development of the development organisation behind StarOffice/OpenOffice is ultimately controlled in reality by Sun. This provides competitive advantages in being able to adapt the development of the format to the development of new features in the programs for word processing etc. in StarOffice/OpenOffice and in the form of best knowledge of new creations, questions of interpretation and so on in relation to the format.

Future Format from Microsoft

Another possible candidate for an XML-based file format for word-processed documents is a future format from Microsoft. Representatives of Microsoft in Denmark have indicated to the working group that Microsoft's strategy is to compete on parameters such as functionality and quality, and not with the aid of closed formats, and that the firm is working on the development of a 100% open, XML-based format that can be used to save Word documents, in other words to be an alternative to the doc format. According to information from Microsoft, the idea is for the new format to be ready for use in conjunction with Office 11, which is due to succeed the current version of Microsoft's office package, Office XP.

The working group has tried to obtain more detailed information on the plans for the format. On the present basis, however, it is difficult to assess the planned format, including whether it will be able to contribute to increasing competition between different suppliers in the area of office software. Microsoft are unable to refer to any detailed descriptions of the format or plans for it, or an official commitment to developing such a format. Microsoft's representatives have stated that the idea is to develop a format equivalent to that which exists today for the spreadsheet program Excel. This format can be used to represent the basic content of a spreadsheet, while elements in a spreadsheet that are based on more advanced features in Excel cannot be represented in the format and are therefore lost if the spreadsheet document is saved (only) in the format.

In an assessment of these suggestions from Microsoft in Denmark, it can be said that Microsoft is already working with the XML standards in many areas. To take an example, Microsoft has been chosen together with the firm Accenture as supplier to what is known as the 'Inforstrukturbase project', which is part of the XML project mentioned above. An assessment must also take into account the fact that it is in Microsoft's interest to preserve its dominance in the area of office software, which is a legitimate business aim for Microsoft. It is not necessarily in Microsoft's interest to eliminate the compatibility problems associated with using alternative products.

Infrastructure Software

The infrastructure area with web servers, mail servers, operating systems etc. is characterised in relation to open source as follows:

- The infrastructure is predominantly based on open standards, primarily the ranges of standards developed under the auspices of the Internet Engineering Task Force (IETF) and the Word Wide Web Consortium
- There is a high degree of competition and supplier independence. The infrastructure thus consists of a mixture of open-source and proprietary software, and in the main there are no compatibility problems between these
- There are many open source products in the infrastructure area that are recognised as being of high quality, including some that are market leaders (the web server Apache) and hold very large market shares (the operating systems Linux and FreeBSD)

In selecting software for a task within the area of infrastructure, it is possible in many cases to choose between solutions that are predominantly based on open-source software and solutions that are predominantly based on proprietary software. This is an extremely favourable situation for government and other customers investing in infrastructure, and the competition has also assisted in bringing down the cost of and spreading the infrastructure, and is consequently in a sense essential if it is to be possible to put the plans for e-government on the agenda.

One of the most important parameters government as a customer should emphasise in selecting infrastructure products is security.

Security in the infrastructure for e-government

The security aspect of software for e-government should be given extremely high priority. A high level of security is a prerequisite both directly for operation of the systems, including keeping them running at all times and at low cost, and indirectly for the public's confidence in the systems and therefore willingness to use them.

Security in software for e-government comprises protection against breaches of secrecy in the content of data communication (e.g. the economic circumstances of members of the public and companies and sensitive personal data) and protection against unauthorised access to computers (for example with the consequence of data being destroyed or a website hacked into so that it is defaced). The security requirements for software depend on the function and purpose of the software and are not limited to those mentioned here.

The problem of security in software for e-government is related to the open structure of the Internet. The fact that the Internet constitutes a link between all the computers connected to it – that is to say, many millions of computers throughout the world – on the one hand means that services are readily available and on the other results in vulnerability:

- The advantage of the Internet as an infrastructure for a large number of systems for e-government is that citizens and enterprises have ready access to the Net. It is easy to be connected to the Internet, and there is cheap and standardised software (including Internet browsers) to use it. Individuals and companies will therefore have easy and inexpensive access to services offered via the Internet
- The drawback is that public services offered on the Internet can be made the object of attack from any computer connected to the Internet

Security and complexity of software

It is possible, in principle, to design software that provides a very high degree of security using encryption systems, digital signature, firewalls and other methods and principles. Experience shows, however, that it is very difficult to design secure software. This is evident from the fact that, despite large resources being invested in software development and security normally being given high priority, new security-related bugs are regularly discovered – whether it is open-source or proprietary software.

It can be said from a general perspective that one of the main causes of security bugs is the complexity of the software. The software has many functions, and consists of thousands and in some cases millions of program lines, so that it is not possible for a single person to have an overview of the software, and there is a complex interaction between its various constituent parts.

Software development requires extensive resources in the form of labour for design, programming and testing/fixing. In the case of both open-source and proprietary software, it is essential that such resources are actually used if a high degree of security is to be attained. When we talk here about development, it is meant in the broadest sense, both the first finished version of a software product and the enhancement (or maintenance) of the product in the form of the work on creating new versions with more features and so on.

We divide the discussion of the various conditions for allocating resources for the development of secure software in open source and traditional projects into two parts: the significance of the available source text (which invites reviews) and what may be termed the development model, where a characteristic feature in open source development is often the participation of many (but where there are no resources for reviews).

Security and availability of source text

The advantage of open-source software in relation to security is the availability of the source code. The advantage consists in the fact that it is possible for everyone to analyse the software, both at an overall architectural level and with regard to the individual parts of the source code. This provides great potential to identify and eliminate bugs. It also enables the user to obtain credible, independent assessments of the software. This also applies to the assessment of a claim regarding a current, disputed bug, as it will be difficult for a supplier to deny a bug if this is documented with reference to the source text.

The publication of source text for security-related software obviously means that the source code is also available to potential hackers and so on. It is a generally recognised principle, however, that security in software must not be conditional on the source text of the software being kept secret, the encryption methods used etc. The secret element must only consist of encryption keys, passwords and similar items. Rules and guidelines for use of the systems must ensure that the various keys are only known to relevant people or programs. If, on the other hand, security is conditional on the actual method used or its implementation being kept secret, unacceptable dependence on such secrets being preserved is introduced, for example a dependence on individual people and organisations who have been involved in the development of the software or who are involved in its operation and maintenance.

Examples of publicly known security-related algorithms and methods are the principles of public key systems (Public Key Infrastructure) and the DES algorithm (Data Encryption Standard). A system based on public keys has to constitute one of the basic elements of e-government. The DES algorithm has been widely used for the encryption of data communication, in recent years in a variant (also publicly known) under the name of triple-DES. The American National Institute of Standards and Technology (NIST) has selected a new algorithm by the name of Rijndal as the successor to DES and triple-DES. The Rijndal algorithm has been published, and NIST has selected it among a number of candidates, the advantages and drawbacks of which have been discussed by security experts quite openly. In the case of all these methods, the method itself is public, and the secret element consists only of the encryption keys.

In the same way that the publication of an encryption method is not a guarantee of the method being good, but is merely an element in a process in which the method is illustrated, publication of source text for security-related software is not in any way a guarantee of security.

Although the security of a particular software product must not be conditional on the methods used and the actual implementation in source code being kept secret, publication may make it easier for a hacker to find the weak spots in the software. The ideal, but impossible situation, is for the source code to be available to those who want to analyse it to fix bugs, and unavailable to those who want to exploit the bugs.

The overall impression of the working group is that where open-source software that has been widely used for several years and has been the object of extensive analysis is involved, and if this process has led to positive assessments by independent experts, this provides the user with an extra degree of security in comparison with proprietary software. The supplier of traditional software can provide selected, independent experts and institutions with access to the source text. There may be increased confidence in the software, but in the nature of things, the process will always be limited to covering a selected circle, and it is not general.

Security: simple bugs versus subtle bugs

In the model for proprietary software, the sale of licences for the software can finance the allocation of resources to systematic testing and fixing of bugs. This raises the question of what mechanisms exist in the open source model to promote allocation of the necessary resources in the form of manpower for development, testing etc.

Custom Built Software

Custom built software covers a wide range, and encompasses for instance:

- highly standardised systems that fulfil a specific function (pay, financial control etc.), for which there is a need in many public organisations, regardless of their administrative tasks
- systems that cover the same needs within a group of public institutions of the same type (e.g. electronic patient records in the healthcare sector)- here called function specific systems.
- highly specialised functions that exist in very few places or in only one place (e.g. the Central Customs and Tax Administration, the central population register)

The characteristic feature of custom built software is that it does not exist in advance, but is developed by a supplier on the basis of a specification of requirements in a public invitation to tender. If the software is to be supplied as open source, this is to form part of the specification of requirements.

Custom built software as open source

The fundamental consideration in connection with requiring custom built software to be supplied as open source will in practice be whether the requirement entails an increase in costs, because the supplier loses the option of selling licences for the same software to other supplier. The requirement to supply custom built software as open source need not be an either/or situation but can be specified as being concerned with the adjustments a supplier makes to existing standard modules, if these are proprietary.

The requirements of the public sector for rights to software for the development of which it has paid can also be varied so that the public organisation has access to the source code and can modify it (which provides better prospects of being able to compete on tasks relating to the maintenance of software) but does not have access to distribution of the software to other organisations with the same software needs (so that the supplier does not lose the option of selling licences to these organisations).

Economic Analyses of Open Source

From the economic point of view, choice of software is equivalent to other types of investments, in that the acquired asset provides a return by creating value over a longer period. At the same time, acquisition of a program is an option that is irreversible, uncertain and takes place in a world undergoing significant technological development, which may lead to the emergence of far better programs. The economic analysis is based on differences between open-source software and proprietary software consisting firstly in differences in the economics of right of use and secondly in differences in the economics of the way in which the software is used, including program development and maintenance, referred to in brief as development. To be able to draw up relevant models for an investment of this kind, it is necessary first to clarify features of software as a financial asset.

The financial investment in software is to a great extent tied to the right of use, which entails a number of financial differences between open-source and proprietary software. This is due to the special circumstances in the production of software. Firstly, programs do not have any physical substance, and consist solely of information, which also means that original and copy are concepts devoid of meaning. Programs can be stored on various media, and the choice of medium is a matter of habit, technology and economics. This also means that there is no physical wear on the software during its use, and there is therefore no physical limit on its useful life. All in all, this means that programs do not have any value as physical products, but have value in the right to use the product.

Secondly, the development costs for software are quite considerable, whereas the marginal costs in copying and distribution are insignificant, particularly where only sale of the right of use is concerned. Particularly for suppliers of software with a large number of units sold, this means that the supplier can set a price that bears no relation to the costs of development and is solely based on the value of the right of use. As different users have different needs, this means that the ideal price is based on the customer's willingness to pay and that software is therefore sold with sharp differentiation of price. The price level may vary over time and in different markets and may show opposite trends, i.e. while the price rises in some markets it falls in others due to competitive considerations. Only strong competition between alternative products can ensure that a downward price trend can become established.

The economics in the sale of software do not follow a classic pattern, because the profit on the last unit sold is largely equal to the sale price. At the same time, the producer does not have any capacity limits, and the size of the market constitutes the only limit on sales. There will therefore be a trend towards the formation of monopolies in standard software, and it will therefore be difficult for competition to be maintained by allowing market forces to dictate development.

Finally, there have been many innovations in software products, which have redefined the prospects of competition in both hardware and software markets, network technologies, WWW etc. Innovations have undermined dominant market positions in some – but not all – cases by creating new competitive conditions.

Organisational circumstances have a strong bearing on efficiency in the utilisation of programs. Where differences between open-source software and proprietary software lead to essentially different organisational requirements for maintenance and updates etc., it is difficult to state the economic differences comparably. We have chosen to focus on examples where we can look at the economics of functionally comparable programs. It is also assumed that programs are acquired on a comparable scale and on a given platform, which means that we look exclusively at software. We therefore avoid entering into a discussion on alternative architectures, while being fully aware that such a choice may have far greater economic consequences than the choice between open-source and proprietary software. Similarly, we assume in our examples that quality is comparable, i.e. we assume that the users consider their needs to be met to an equivalent extent.

Unlimited access to using copies of an open-source software program implies an immediate benefit for users of open-source software, as the costs of right of use are independent of the number of users, when compared with proprietary software, where the right of use is paid for as a sum per user, although licence price often falls with the number of users. At the same time, the rights also lay down the options for making changes to software or integrating open-source software with proprietary or customer-owned software. When comparable products exist, this rights perspective therefore points to direct and clear economic advantages in acquiring open-source software rather than proprietary software products.

The formulation of rights also has an impact on economic differences in the longer-term perspective. This relates, for example, to free choice of supplier for maintenance, and in certain circumstances even free access to maintenance without payment, adaptation to future program procurements, free access to choice of supplier of integration software and so on.

When there are no directly comparable products, the economics of the design of open-source software and proprietary software depends on how specifically the product is designed. A software product that can only be used by the customer concerned ('organisation-specific software') cannot justify price differences between open-source and proprietary software, as all the development costs have to be covered by this one sale. An open source licence for a software product with several/many potential users will mean a potential loss of earnings for the developer. This applies both to specific software and to general software such as infrastructure software or desktop software. This could directly justify a significantly higher procurement price in relation to the first customer, but competition for the contract will reduce a difference of this kind, as the winner of the contract will be left with software skills of value, including those for potential customers for the product, and with relatively better prospects of gaining the development contracts (installation, adaptation, enhancement) that are essential if efficient organisational utilisation is to be achieved.

The extent of the economic differences between open-source and proprietary software depends on other factors in addition to those discussed above, as prices and terms of licence for proprietary software are altered and adapted to market conditions, so that differences between open-source and proprietary software vary over time under the influence of market conditions.

Suppliers with their own software products have responded in widely differing ways to the penetration of their markets by open-source software products. Responses range from recognition and acceptance, with the inclusion of their own resources to maintain and develop a product, to pure rejection and suspicion. There are wide differences in the price strategies adopted by different competitors in relation to open-source software, depending on market shares and potential for the individual open-source software products. In other words, we cannot talk of open-source software as a family of products where exactly the same economic advantages (and drawbacks) apply to all products, as these products have been viewed in the light of the alternative options among proprietary software on the market.

The Economic Development Perspective

If the right of use is to be of value, the software has to be implemented, procedures changed, users trained and so on. Both this short-term change in the organisation and the more long-term change in the organisation in conjunction with renewal and replacement of software form the basis for the analysis of differences between open-source software and proprietary software. This is referred to overall as the development perspective.

For proprietary software, development and maintenance are only done by the owner of the copyright to the program. Responsibility for development and maintenance for open-source software may rest with both the original holder of rights (who stipulates that open source rights are to apply from then on) and the individual user, as any user in principle can enhance the product and customise it to his or her own wishes, while complying with the open source rights that apply in this situation. User-as-developer-and-tester is the decisive difference from proprietary software, as open-source software via user groups on the Internet has the potential for rapid and versatile, mutual support with free development of software improvements and corrections, i.e. without obstructive software rights.

The development perspective covers both centralised and highly decentralised development concepts for open-source software products, which affects the time perspective, skills requirements and organisation. The time perspective for an economic asset is of decisive importance to its value. It has been noted that for every Danish krone spent on acquisition of IT, ten kroner is spent on organisational adaptation, development etc. The economic life-span of software must therefore not just be compared with the procurement price. When the other costs are added, the requirement for software life increases to include cost-effectiveness, in the same way that replacing existing software makes greater demands on software product improvements than are equivalent to the additional price for new purchase. Where these improvements and their (expected) productivity effects cannot be documented, a change for other reasons must therefore of necessity be linked to a very high degree of functional and user-interface agreement with the software used to date in order to minimise installation, maintenance and training costs. These other reasons may be a revision necessitated by bugs and deficiencies in the software.

The difference between internal IT skills being available and external consultants with relevant skills having to be called in gives rise to differences in the structure of the operating costs as well as that of the organisation, which may make comparisons between solutions in the short term difficult. It is only reasonable to compare different solutions if a multi-year perspective is adopted.

Disregarding the development perspective would mean overlooking the very substantial difference between open-source software and proprietary software: unobstructed entitlement to adapt and expand open-source software products locally in an organisation, regionally in an authority area or nationally for a public institution. The next section describes software investments as options.

Options

Options form the key to understanding the economics in the use of software in a development perspective, reflecting the fact that an investment has to be based on the principle that investments are generally irrevocable, uncertain, and have timing problems, so that it is necessary to bring the time of the investment into the overall considerations. In the case of real options, a project (the investment) is identified and a decision-maker assesses how its premises and potential are developing before the investment is made, while the classic method is a financial assessment comparing the expected return from a project with that from financial securities. Software investments as a rule entail follow-on investments, the value of which is significant for the original investment value. Upgrades can be regarded as a follow-on investment where purchasers of previous versions obtain the new version more cheaply than new purchasers. We shall look at three aspects of software as real options: irreversibility, uncertainty and upgrading.

Irreversible investments are investments that cannot be cancelled without financial loss. Investments of this kind therefore have far greater consequences than those linked to the narrower rights perspective. The degree of irreversibility of an investment can be assessed on the basis of the value of the real option expressed as 'switching costs'.

The second factor in real options is uncertainty. In relation to software, this factor applies constantly during the time a software product is in use, as a whole series of requirements of development opportunities have to be met for the acquired software to fulfil the expectations on productivity gains. Technological innovations can reduce the value of the acquired software if new software offers substantially improved functions. New requirements for interaction with other software can reduce the functionality of previously acquired software. There will not be certainty about any of these changes in the procurement of software, but the history of software shows that there will be technological innovations. By offering integrated software products, the market has tried to reduce uncertainty over the functional value of the product, but has increased the requirement regarding organisational adaptation to the product concerned and in so doing has increased the organisational uncertainty over whether the organisation is capable of harvesting the functional value.

Upgrading is a special problem in assessing software as an investment. From the point of view of the investor, the economic life-span is limited by the emergence of better software, where added value of the new software is greater than with the present software. In other words, software is not to be replaced until the time when the value of future use (including purchase) of a new program is greater than the value of continued use of the existing one. From the producer's point of view, upgrading is an economic necessity, as sales opportunities for unchanged software are expected to be reduced for every licence sold (not because of pirate copying but because of declining novelty value and the emergence of competing products). The producer therefore has a strong incentive to continue to offer new facilities. The greatest competition, particularly for products with great market coverage, comes from earlier versions of the same product.

The option elements of software products depend upon the moment of investment; external influence like changing software technology as well as internal IT organisational capacity and competences vary over time. What is a good time and a good choice of software are determined by circumstances that are uncertain, because options are an assessment of future circumstances that are not only under the investor's own control. A more detailed account will be given below of the costs associated with software in a development perspective, involving the options perspective that has been outlined above.

Open Source Governance

Open source governance is a political philosophy which advocates the application of the philosophies of the open source and open content movements to democratic principles in order to enable any interested citizen to add to the creation of policy. Legislation is democratically opened to the general citizenry in this way, allowing policy development to benefit from the collected wisdom of the people as a whole.

Some envision this form of governance as a post-national "virtual state" governing structure, where policy-setting is decoupled from territorial management. Some models are significantly more sophisticated than a wiki, incorporating levels of control or scoring to mediate disputes. In any event, the idea demonstrates the still untapped potential of how open source philosophies can merge with government.

A typical core principle is the concept of a "central codebase" in the form of a set of policies that are maintained in a public registry and that are *infinitely reproducible*. "Distributions" of this policy-base are released (periodically or dynamically) for use in localities, which can apply "patches" to customize them for their own use. Localities are also able to cease subscribing to the central policy-base and "fork" it or adopt someone else's policy-base. In effect, the government stems from emergent cooperation and self-correction among members of a community. As the policies are put into practice in a number of localities, problems and issues are identified and solved, and where appropriate communicated back to the core.

Because so much information must be gathered for the overall decision-making process to succeed, however, technology may provide important forces leading to the type of empowerment needed for participation in this kind of government, especially those technological tools that enable community narratives and correspond to the accretion of knowledge.

The open politics theory combines aspects of the free software and open content movements, promoting decision making methods claimed to be more open, less antagonistic, and more capable of determining what is in the public interest with respect to public policy issues.

While it can be confused with the newer idea of open source politics, open politics is not so much a movement as a theory based on participatory democracy and deliberative democracy, informed by e-democracy and netroots experiments, applying argumentation framework for issue-based argument as they evolved in academic and military use through the 1980s to present. Some variants of it draw on the theory of scientific method and market methods, including prediction markets and anticipatory democracy.

Criteria

Open politics can be reduced to a list of criteria:

1) Anyone can participate, including anonymously
2) All participants are equals, and resolve disputes via equal power relationships
3) All actions are transparent, and no one has more power to review them than anyone else
4) All contributions are recorded and preserved, and these records cannot be altered
5) All deliberation is structured, or can be put in structured form to resolve disputes
6) All content is re/organized and refactored by participants
7) Partisan behavior is limited by the format, rules set by factions themselves, and laws extant in the society or community which will be affected by the political decision
8) Control of the forum can, at least in theory, pass to the most trusted users, not the ones who started the forum

Some experts apply strict criteria of democracy, rootedness, legality, equality of access, and even ecological integrity, so as to ensure that there are absolutely no rights lost in moving polity into an online arena. In other words, they wish to expand participation to mobile and remote persons, including disadvantaged ones, and undo some of the inequities inherent in using electronic media.

Underlying Preferences and Ideals

Underlying these criteria in turn are ideals and preferences that resemble those of other democratic political movements:

1. decentralization of authority: giving the widest and most potent franchise to citizens is thought to minimize what economists call the principal-agent problem, or the tendency for managers to abuse authority.

2. centralization of information: the use of information technology to facilitate communication challenges is key to the practicality of the process.
3. equality of opportunity: anyone can participate in deliberation, with the expectation that people themselves select to participate on issues in which they have the greatest stake, expertise or both. Open politics treats the expert and the citizen as equals, implying that the experts are obliged to convince the citizens directly, rather than using representatives as intermediaries/brokers of policy. This use of peer review is emphasized as the best method to determine what is true or good (with the understanding that this should change over time).
4. encouraging diversity of thought, such that multiple positions and arguments are created, refined and compared; usually the more the better, provided they are succinct.

Some theorists describe the ideals as similar to libertarian and green politics with the emphasis on peer review and scientific method within political science. However, the idea that political science could apply falsificationism is controversial, and despite an invitation to contradict and counter arguments, the rigorous application of scientific method is not part of every open politics service.

Open politics theory grew from earlier work in online deliberation and deliberative democracy, which in turn drew on research in issue-based argument and early hypertext and computer-supported collaboration research of the early 1980s.

The "Imagine Halifax" project was designed to create a citizens' forum for elections in Halifax, Nova Scotia in fall 2004. Founded by the widow of the late Tooker Gomberg, a notable advocate of combining direct action with open politics methods, IH brought a few dozen activists together to compile a platform (using live meetings and email and seedwiki followup). When it became clear that candidates could not all endorse all elements of the platform, it was then turned into questions for candidates in the election. The best ideas from candidates were combined with the best from activists—the final scores reflected a combination of convergence and originality. In contrast to most such questionnaires, it was easier for candidates to excel by contributing original thought than by simply agreeing. One high scorer, Andrew Younger, had not been involved with the project originally but was elected and appeared on TV with project leader Martin Willison. The project had not only changed its original goal from a partisan platform to a citizen questionnaire, but it had recruited a previously uninvolved candidate to its cause during the election. A key output of this effort was a glossary of about 100 keywords relevant to municipal laws.

The 2004–05 Green Party of Canada Living Platform was a much more planned and designed effort at open politics. As it prepared itself for an electoral breakthrough in the 2004 federal election, the Green Party of Canada began to compile citizen, member and expert opinions in preparation of its platform. During the election, it gathered input even from Internet trolls including supporters of other parties, with no major problems: anonymity was respected and comments remained intact if they were within the terms of use at all. Despite, or perhaps because of, its early success, it was derailed by Jim Harris (politician), the party's leader, when he discovered that it was a threat to his status as a party boss. The Living Platform split off as another service entirely out of GPC control and eventually evolved into openpolitics.ca and a service to promote wiki usage among citizens and political groups.

The Liberal Party of Canada also attempted a deep policy renewal effort in conjunction with its leadership race in 2006. While candidates in that race, notably Carolyn Bennett, Stéphane Dion and Michael Ignatieff, all made efforts to facilitate web-threaded policy-driven conversations between supporters, all failed to create lateral relationships and thus also failed to contribute much to the policy renewal effort.

Direct Democracy

Direct democracy, classically termed pure democracy, is a form of democracy and a theory of civics in which sovereignty is lodged in the assembly of all citizens who choose to participate. Depending on the particular system, this assembly might pass executive motions, make laws, elect or dismiss officials, and conduct trials. Direct democracy stands in contrast to representative democracy, where sovereignty is exercised by a subset of the people, usually on the basis of election. Deliberative democracy incorporates elements of both direct democracy and representative democracy.

Some anarchists (usually social anarchists) have advocated forms of direct democracy as an alternative to the centralized state and capitalism;

however, others (such as individualist anarchists) have criticized direct democracy and democracy in general for ignoring the rights of the minority, and instead have advocated a form of consensus decision-making. Libertarian Marxists, however, fully support direct democracy in the form of the proletarian republic and see majority rule and citizen participation as virtues. The Young Communist League, USA in particular refers to representative democracy as "bourgeois democracy," implying that they see direct democracy as "true democracy

Double majorities are, first, the approval by a majority of those voting, and, second, a majority of cantons in which a majority of those voting approve the ballot measure. A citizen-proposed law (i.e. initiative) cannot be passed in Switzerland at the national level if a majority of the people approve, but a majority of the cantons disapprove. For referendums or propositions in general terms (like the principle of a general revision of the Constitution), the majority of those voting is enough.

In 1890, when the provisions for Swiss national citizen lawmaking were being debated by civil society and government, the Swiss copied the idea of double majorities from the United States Congress, in which House votes were to represent the people and Senate votes were to represent the states. According to its supporters, this "legitimacy-rich" approach to national citizen lawmaking has been very successful. Kobach claims that Switzerland has had tandem successes both socially and economically which are matched by only a few other nations, and that the United States is not one of them. Kobach states at the end of his book, "Too often, observers deem Switzerland an oddity among political systems. It is more appropriate to regard it as a pioneer." Finally, the Swiss political system, including its direct democratic devices in a multi-level governance context, becomes increasingly interesting for scholars of European Union integration.

Democratic Schools

A democratic school is a school that centers on providing a democratic educational environment featuring "full and equal" participation from students and staff. These learning environments position youth voice as the central actor in the educative process by engaging students in every facet of school operations, including learning, teaching, leadership, justice, and democracy, through experience.Adult staff support students by offering facilitation according to students' interests.

Sudbury model of democratic education schools are run by a School Meeting where the students and staff participate exclusively and equally. Everyone who wishes to attend can vote, and there are no proxies. As with direct democracy elsewhere, when an issue is important to a person, that person participates in the meeting where the issue is to be discussed. Otherwise, they usually do not bother.

Another tenet of Sudbury schools is giving students the power to choose what to do with their time: individual freedom, freedom of choice, and learning through experience. There are no required classes, and usually there is no requirement to take classes at all. Students are free to choose an activity that they desire, or feel the need to do. They are free to continue activities for as long or short a time as they see fit. In this way, they learn self-discipline, self-regulated learning and self-initiation. They also learn faster and retain more knowledge, thanks to the engagement in activities that they are passionate about. The students at these schools are responsible for and empowered to direct their own education from a very young age. All this is facilitated, supported and managed by the school's democratic framework.

Parliamentary Informatics

Parliamentary informatics is the application of information technology to the documentation of legislative activity. The principal areas of concern are the provision, in a form conveniently readable to humans or machines, of information and statistics about:

1) individual legislators
2) particular legislative proposals
3) votes thereon
4) text of legislation

Parliamentary informatics is carried on both by officials of legislatures and by private for-profit and non-profit actors, with motivations ranging from the administration of parliaments to lobbying and facilitating democratic discourse.

The division of activities between official and unofficial activity differs widely between polities, even within a single country. There exists substantial overlap with disciplines such as psephology and, as far as the text of successfully enacted legislation is concerned, legal informatics in general.

Websites of Various Countries

Access to parliamentary information is the most significant issue in parliamentary informatics; access may be inhibited technologically through the provision of parliamentary information in a form not amenable to processing by machines, or legally, by copyright or other protection of the parliamentary record.

Ultimately, it may be possible to provide citizens with a comprehensive overview of all the legislation going through parliaments everywhere in the world, which would allow the tracking of the implementation of international treaties and of general legislative trends.

The website http://www.undemocracy.com/ gives hyperlinked access to transcripts of the General Assembly and Security Council of the United Nations, with parsed voting records.

Une Votoj (in English, "I vote") is a website that provides information about the Parliament of Albania. The site structures Parliamentary plenary and committee sessions and agendas, and summaries of the proceedings. It also provides biographical notes about elected members, attendance and voting records.

OpenAustralia.org is a community-funded site which was launched in June 2008 and automatically collects and publishes in an easy-to-read format the transcripts of all Australian Government House of Representatives and Senate debates, questions, discussions and notices. OpenAustralia.org also made available the Register of Senators' and Members' Interests online for the first time ever in Australia. The site was developed in Australia with the assistance of the UK-based mySociety group who also built TheyWorkForYou from which OpenAustralia has been adapted. The site allows comment and discussion against debates and allows users to track the level of participation of Member of the House in debates, including statistics on their appearances in the House — so citizens can keep track of their representatives and be fully informed and involved with their Member's performance in the House. Within weeks of being launched OpenAustralia attracted a lot of attention and positive feedback from both the public and government and is set to pave the way for closer working relationships between government and citizens.

The National Council of Austria is covered by meinparlament.at. The website was modeled after a German partner website.

"Africa i-Parliaments" is the portal of the regional initiatives of the United Nations Department of Economic and Social Affairs (UN/DESA) aimed at strengthening the role of African Parliaments in fostering democracy and good governance by developing common information services and tools, and building information management capabilities with the objective of making Parliaments i-nterconnected, i-nformed i-ndependent, or in short, i-Parliaments.

The project includes two main initiatives: AKOMA NTOSO (Architecture for Knowledge-Oriented Management of African Normative Texts using Open Standards and Ontologies) proposes an XML document schema providing sophisticated description possibilities for several Parliamentary document types (including bills, acts and parliamentary records, etc.). It aims to standardize simple, technology-neutral representations of Parliamentary Documents in order to improve inter-Parliamentary cooperation and reduce the costs of Parliamentary IT support systems. AKOMA NTOSO defines a set of recommendations and guidelines for e-Parliament services in a Pan-African context and provides an enabling framework for the effective exchange of machine readable Parliamentary Documents such as legislation, debate record, minutes, etc.

Bungeni – Parliamentary Information System is an end-to-end suite of applications that provides a world-leading solution for drafting, managing, consolidating and publishing legislative and other parliamentary documents. Bungeni aims to increase the efficiency of parliamentary activities and make Parliaments more open and accessible to citizens—virtually allowing them "inside the Parliament" or Bungeni, the Kiswahili word for "inside the Parliament".

The Parliament website (Folketinget) has records of votes, speeches, laws and written questions in plain form. Folkets Ting is a volunteer-run website with the same data that also offers rss feeds and lets users comment on content. "Folkets Ting" means "Peoples Parliament" and is a pun on "Folketinget". HvemStemmerHvad has elaborate parliament voting statistics and tracks media mentions of politicians.

www.votewatch.eu has collected and displays full records of the activities of the Members of the European Parliament starting with 2004. VoteWatch.eu is based on know-how from Romanian IPP's website (which has already been shut down) and uses the expertize of academics from London School of Economics and Universite Libre de Bruxelles. The website displays full records of voting, attendance and other activities (parliamentary questions, speeches, reports etc.) of MEPs from all 27 member states. Furthermore, it shows a set of statistics that provide detailed information on how the coalitions are forming between the various European Political Groups. VoteWatch.eu has been updated for the 2009-2014 parliamentary term.

Testbeeld Europa (Dutch language) shows how the Dutch Members of the European Parliament voted. Made by the nonpartisan, non-profit Instituut voor Publiek en Politiek (IPP).

IPP in Romania ran a full site monitoring MEPs. It shut down in December 2008.

Czech-Slovak KohoVolit.eu (in English, too) prepares calculators on match between user and political groups and parties in EP, and individual MEPs. It is available for the 2009- term.

Follow the Money which aims to monitor the EU budget and runs the twin sites FarmSubsidy and FishSubsidy

An association of free software campaigners known as April has been gathering declarations from candidates standing for election about their opinion on the issue of free software, and providing an interface for voters to look up the results.

1. Since 2006 Mémoire Politique publishes a toolbox designed to help you reach members of French and European Parliaments, and track their voting records.
2. The 14th of septembre 2009, NosDeputes.fr is released. It tracks the activity of French Members of the Parliament (Assemblée Nationale). Their data are released under open licences (GPL and CC-by-sa). It is hosted by Regards Citoyens.
3. Mon-Depute.fr was the first site that publishes MP's voting rolls.

The German federal parliament Deutscher Bundestag and two state parliaments (Landtag) are covered by abgeordnetenwatch.de. The website supports the transparency of MP's voting behavior and MP's supplementary income. However, the main focus is on public Q&A. Citizens can ask questions, which can be answered by members of parliament. Newer projects like bundestagger.de, which tries to make parliamentary protocols accessible, are parts of the newly founded OpenData Network, a group of software developers and political activists.

Georgian Parliament has information on MPs, and their voting records. Civilin has information about legislation. My Parliament aggregates it all together.

KildareStreet.com (named for Kildare Street, where the Irish Parliament is located) is a volunteer-made site launched in April 2009 which collects and republishes the transcripts of debates in Dáil Éireann and Seanad Éireann,and parliamentary questions. It, like the Australian project OpenAustralia is partly derived from the code base of TheyWorkForYou in the UK. It intends to expand to cover members' expenses and register of interests in the future.

Openparlamento allows to track legislative progress, votes and Parliament members using official data from italian Parliament. The platform lets users comment, vote and emend on every parliament act, supplies official legislative text and rss feeds. openparlamento is a non partisan project realised by openpolis.

The Jordanian Parliament Monitor http://www.jpm.jo/ is a website that provides information about the Parliament of Jordan. The site presents profiles of all members of Parliament, information about the MP's deeds and stances on public policy issues, bills and law proposals, questions to the Cabinet, interventions during plenary and committee sessions. The website also includes news about the various parties and blocs represented in Parliament, statistics about the Parliament and about MP's performance as well as the studies and reports published by the Monitor. It was established in 2009, and is run by [Al-Quds Center for Political Studies] http://alqudscenter.org/English/, a Jordanian think-tank based in Amman.

Mzalendo: Eye On Kenyan Parliament provides electronic information about the Kenyan parliament, which is otherwise only available offline.

KDI Kosovo provides electronic reports and voting records from Kosovo's parliament.

The Lebanese Parliamentary Monitor is a website that provides information about the Parliament of Lebanon. The site presents profiles of all 128 Lebanese members of Parliament, information about the MP's deeds and stances on 50 public policy issues, bills and law proposals, questions to the Cabinet, interventions during plenary and committee sessions and law challenges submitted to the Lebanese Constitutional Council. The website also includes statistics about the Parliament and about MP's performance as well as the monthly newsletters of the Monitor, summarizing the parliamentary activities. The Monitor organizes regular public hearings gathering citizens and one or two MPs from a certain district, to debate the MPs' commitment to voicing the concerns of aspirations of their constituents. It was established in 2006, and is run by Nahwa al-Muwatiniya(Towards Citizenship), a Beirut-based non-governmental civil society organization.

Atviras Seimas (Open Parliament) provides statistics for MP attendance, votes, speeches, rebellions, travel maps, popularity ratings based on internet search result counts. It is a volunteer-run website.

Mano Seimas (My Parliament) publishes voting records for interesting or controversial bills, provides MP's biography and allows to ask questions and receive answers from MPs. Part of the published data is fetched through API provided by Atviras Seimas. Mano Seimas is a part of e-democracy project run by Institute of International Relations and Political Science, Vilnius University.Manobalsas.lt (myvote) is a smart voting tool allowing users to take a test and find out which political party and candidate best matches their opinion. The site launched in 2008 before National parliament elections. More than 300 candidates registered on the site.

TheyWorkForYou.co.nz hosts a user friendly version of the NZ Parliament debate transcripts, and summarizies activity by bill, by ministerial portfolio and by organisations making submissions to the Parliament. It also provides lists of how parties voted on bills in Parliament. It is a is a volunteer-run project.

CommoNZ provides lists of how MPs votes on non-party votes in Parliament (in New Zealand, many votes are formally conducted by the parties rather than the individuals, even in respect of MPs with constituencies).

wiekiesjij.nl helps you find the politician of your choice based on expertise, professional background, age, religious affiliation, and political experience. More than 90% of Dutch candidates for national parliament filled-in an online questionnaire on which the system was based. It was made by the foundation "The New Way of Voting" (See: Het Nieuwe Stemmen or its international forum: Forum

Stemmentracker was launched in the national elections in November 2006. For crucial votes the site provides an introduction to the vote and shows how parties voted. A voter can also vote, so the systems shows the parties that voted most like the user. Made by the nonpartisan, non-profit Instituut voor Publiek en Politiek (IPP). IPP looking for possibilities for keeping it up to date in the future.

The Institutul pentru Politici Publice has created a web site with lots of structured data about the Romanian Parliament. Adrian Moraru, deputy director with IPP said in an interview with mySociety explains how they have had to obtain much of their Parliamentary data by court action. Currently the site is offline because of discontinued funding, tough IPP plans a new version to be launched in 2009.

KohoVolit.eu is a volunteer-run website providing a sample of votings of the Slovak Parliament allowing the user to test her or his own preferences against the MPs.

The situation in the United Kingdom is very active, with several of the major volunteer projects run by mySociety. The official UK Parliament website provides transcripts of the Parliamentary debates and votes in plain text form, and these are parsed by a project known as parlparse into a timeline of publicly available structured XML files. These files provide the data for TheyWorkForYou, which hosts the speeches in a user friendly form and creates email alerts and rss feeds, and Public Whip, which keeps track of the votes and allows for an expression of their meanings in plain English.

Independently of this, the academic Philip Cowley at Nottingham University researches specifically into how MPs vote through his Revolts website and publications.

Community Informatics

Community informatics refers to an emerging field of investigation concerned with principles and practices related to information and communication technology (ICT) with a focus on the personal, social, cultural or economic development of and within communities or groups. It is located within the academic discipline of Information Systems and draws on insights on community development from other social sciences disciplines. It is a cross- or interdisciplinary approach interested in the utilization of ICTs for different forms of community action, as distinct from pure academic study or research about ICT effects

Human activity, with rare exceptions, is lived in communities. The concept of "community" and its connections to different forms of social networks has seen countless analyses and critiques. It can be seen in two contexts. First, the reality of "community" as a lived and working experience where the lived community and the physical community overlap (as for example in rural areas). Second, "community" as a concept is applied (as for example in urban areas or in virtual networks) to attempts to enable (from a practitioner perspective) and explore (from a research perspective) the reality and significance of neighbourhoods, ethnic and cultural associations, and professional interests among others to provide frameworks for social meaning and social action. Thus "communities", as people coming together in pursuit of their common aims or shared practices both physically and electronically-enabled, proliferate even while their "researched" reality remains in considerable dispute.

To date there has been very considerable investment in supporting the electronic development of business communities, one-to-many social tools (for example, corporate intranets, or purpose-built exchange and social networking services such as Ebay, or Myspace), or in developing applications for individual use. There is far less understanding, or investment in human-technical networks and processes that are intended to deliberately result in social change or community change, particularly in communities for whom electronic communication is of second order interest to having an adequate income or social survival.

The communal dimension (and focus of Community Informatics) results in a strong interest in studying and developing strategies for how ICTs can enable and empower those living in physical communities. This is particularly the case in those communities where ICT access is done communally as for, example through Telecentres, information kiosks, Community Multimedia Centres, and others. This latter set of approaches has become of very considerable interest as Information and Communications Technology for Development (ICT4D) has emerged as significant element in strategic (and funding) approaches to social and economic development in Less Developed Countries. ICT4D initiatives have been undertaken by public, NGO and private sector agencies concerned with development such as the United Nations Development Program, the World Bank, the Swiss Agency for Development and Cooperation (SDC), the MS Swaminathan Research Foundation; have emerged as a key element in the poverty alleviation component of the UN's Millennium Development Goals; and as important directions for private sector investment both from a market perspective (cf. the "Bottom of the Pyramid") and from companies concerned with finding a delivery channel for goods and services into rural and low income communities.

There is thus growing interest in Community Informatics as an approach to the understanding of how different information and communication technologies can enable and empower ordinary, and deprived social and physical communities in relation to the achievement of their collective goals.

Conceptual Approaches

As an academic discipline community informatics (CI) can be seen as a field of practice in applied information and communications technology. The term was brought to prominence by Michael Gurstein. He brought out the first representative collection of academic papers, although others, such as Brian Loader and his colleagues at the University of Teesside used the term in the mid-1990s.

CI brings together the practices of community development and organization, and insights from fields such as sociology, planning, computer science, critical theory, women's studies, library and information sciences, management information systems, and management studies. Its outcomes-community networks and community-based ICT-enabled service applications-are of increasing interest to grassroots organizations, NGOs and civil society, governments, the private sector, and multilateral agencies among others. Self-organized community initiatives of all varieties, from different countries, are concerned with ways to harness ICT for social capital, poverty alleviation and for the empowerment of the "local" in relation to its larger economic, political and social environments. Some claim it is potentially a form of 'radical practice'.

Community informatics may in fact, not gel as a single field within the academy, akin for example to Information Systems or Management Information Systems, but remain a convenient locale for interdisciplinary activity, drawing upon many fields of social practice and endeavour, as well as knowledge of community applications of technology. However, one can begin to see the emergence of a postmodern "trans-discipline" presenting a challenge to existing disciplinary "stove-pipes" from the perspectives of the rapidly evolving fields of technology practice, technology change, public policy and commercial interest. Whether or not such a "trans-discipline" can maintain its momentum remains to be seen given the incertitude about the boundaries of such disciplines as community development.

Furthermore, there is a continuing disconnect between those coming from an Information Science perspective for whom social theories, including general theories of organisation are unfamiliar or seemingly irrelevant to solving complex 'technical' problems, and those whose focus is upon the theoretical and practical issues around working with communities for democratic and social change.

Given that many of those most actively involved in early efforts were academics (drawn from a variety of disciplines including Anthropology, Computing Science, Development Studies, Information Science and Systems, Management, Planning, Sociology, and Social Work among others) it is only inevitable that a process of "sense-making" with respect to these efforts would follow on quite quickly from the flurry of "tool-making" efforts. These academics, and some community activists connected globally through the medium.

A first formal meeting of researchers with an academic interest in these initiatives was held in conjunction with the 1999 Global Community Networking Conference in Buenos Aires, Argentina. This meeting began the process of linking Developed Country community based ICT initiatives and research with initiatives being undertaken in Less Developed Countries often as part of larger economic and social development programmes funded by agencies such as the UN Development Programme, World Bank, or the International Development Research Centre. For the first time, the efforts being undertaken in using ICT for economic and social development purposes in the Developed Countries began to find common interests and common cause with parallel efforts in Less Developed Countries and similarly those with academic or research activities in these areas began to see common and overlapping interests. For example, the issue of sustainability as a technical, cultural, and economic problem for community informatics has resulted in a special issue of the Journal of Community Informatics as well as the subject of ongoing conferences in Prato, Italy and other conferences in South Africa.

Social informatics beyond an immediate concern for a community.Social Informatics refers to the body of research and study that examines social aspects of computerization – including the roles of information technology in social and organizational change, the uses of information technologies in social contexts, and the ways that the social organization of information technologies is influenced by social forces and social practices. Historically, social informatics research has been strong in the Scandinavian countries, the UK and Northern Europe. Within North America, the field is represented largely through independent research efforts at a number of diverse institutions. Social informatics research diverges from earlier, deterministic (both social and technological) models for measuring the social impacts of technology. Such technological deterministic models characterized information technologies as tools to be installed and used with a pre-determined set of impacts on society dictated by the technology's stated capabilities. Similarly, the socially deterministic theory represented by some proponents of the social construction of technology (SCOT) or social shaping of technology theory as advocated by Williams & Edge in 1996 see technology as the product of human social forces.

There is a tension between the practice and research ends of the field. To some extent this reflects the gap, familiar from other disciplines such as community development, community organizing and community based research In addition, the difficulty that Information Systems has in recognising the qualitative dimension of technology research means that the kind of approach taken by supporters of community informatics is difficult to justify to a positive field oriented towards solutions of technical, rather than social problems. This is a difficulty also seen in the relationship between strict technology research and management research. Problems in conceptualising and evaluating complex social interventions relying on a technical base are familiar from community health and community education There are long-standing debates about the desire for accountable—especially *quantifiable* and outcome-focused social development, typically practised by government or supported by foundations, and the more participatory, qualitatively-rich, process-driven priorities of grass-roots community activists, familiar from theorists such as Paulo Freire, or Deweyan pragmatism.

Some of the theoretical and practical tensions are also familiar from such disciplines as program evaluation and social policy, and perhaps paradoxically, Management Information Systems, where there is continual debate over the relative virtue and values of different forms of research and action, spread around different understandings of the virtues or otherwise of allegedly "scientific" or "value-free" activity (frequently associated with "responsible" and deterministic public policy philosophies), and contrasted with more interpretive and process driven viewpoints in bottom-up or practice driven activity. Community informatics would in fact probably benefit from closer knowledge of, and relationship to, theorists, practitioners, and evaluators of rigorous qualitative research and practice.

A further concern is the potential for practice to be 'hijacked' by policy or academic agendas, rather than being driven by community goals whether in Developed Country "Digital Divide" programs or in projects situated in Less Developed Countries. The ethics of technology intervention in indigenous or other communities has not been sufficiently explored, even though ICTs are increasingly looked upon as an important tool for social and economic development in such communities.

Moreover, neither explicit theoretical positions nor ideological positioning has yet emerged. Many projects appear to have developed with no particular disciplinary affiliation, arising more directly from policy or practice imperatives to 'do something' with technology as funding opportunities arise or as those at the grassroots (or working with the grassroots) identify ICT as possible resources to respond to local issues, problems or opportunities.

The papers and documented outcomes (as questions or issues for further research or elaboration) on the wiki of the October 2006 Prato conference demonstrate that many of the social, rather than technical issues are key questions of concern to any practitioner in community settings: how to bring about change; the nature of authentic or manufactured community; ethical frameworks; or the politics of community research.

Research and practice ranges from concerns with purely virtual communities; to situations in which virtual or online communication are used to enhance existing communities in urban, rural, or remote geographic locations in developed or developing countries; to applications of ICTs for the range of areas of interest for communities including social and economic development, environmental management, media and "content" production, public management and e-governance among others. A central concern, although one not always realized in practice is with "enabling" or "empowering" communities with ICT that is, ensuring that the technology is available for the community. This further implies an approach to development which is rather more "bottom up" than "top down".

Areas of concern range from small-scale projects in particular communities or organizations which might involve only a handful of people, such as telecentres; an on online community of disabled people; civic networks and to large national, government sponsored networking projects in countries such as Australia and Canada or local community projects such as working with Maori families in New Zealand. The Gates Foundation has been active in supporting public libraries in countries such as Chile. An area of rapidly developing interest is in the use of ICT as a means to enhance citizen engagement as an "e-Governance" counterpart (or counterweight) to transaction oriented initiatives.

A key conceptual element and framing concept for Community Informatics is that of "effective use" introduced initially by Michael Gurstein in a critique of a research pre-occupation with the Digital Divide as ICT "access" CI is concerned with how ICTs are used in practice and not simply facilitating "access" to them and the notion of "effective use" is a bridge between CI research (research and analysis of the constituent elements of effective use), CI policy (developing enabling structures and programmes supportive of "effective use") and practice (implementing applications and services in support of local communities).

Many practitioners would dispute any necessary connection to university research, regarding academic theorising and interventions as constraining or irrelevant to grassroots activity which should be beyond the control of traditional institutions, or simply irrelevant to practical local goals.

Some of the commonalities and differences may be in fact be due to national and cultural differences. For example, the capacity of many North American (and particularly US) universities to engage in service learning as part of progressive charters in communities large and small is part of a long-standing tradition absent elsewhere. The tradition of service learning is almost entirely absent in the UK, Australia, or New Zealand, (and of limited significance in Canada) where the State has traditionally played a much stronger role in the delivery of community services and information.

In some countries such as the UK, there is a tradition of locally based grassroots community technology, for example in Manchester, or in Hebden Bridge. In Italy and the Netherlands, there also appears to have been a strong connection between the development of local civic networks based around a tradition of civic oppositionism, connected into the work of progressive academics.

There are emerging online and personal networks of researchers and practitioners in community informatics and community networking in many countries as well as international groupings. The past decade has also seen conferences in many countries, and there is an emerging literature for theoreticians and practitioners including the on-line Journal of Community Informatics.

References

Androutsellis-Theotokis, et.al. (2010). "Open source software: A survey from 10,000 feet" (PDF). *Foundations and Trends in Technology, Information and Operations Management* 4 (3-4): 187–347.

Berry, D M.& Moss, Giles (2006). *Free and Open-Source Software: Opening and Democratising e-Government's Black Box*. Information Polity Volume 11. (1). pp. 21–34

Chris DiBona and Sam Ockman and Mark Stone, ed. (1999). *Open Sources: Voices from the Open Source Revolution*. O'Reilly.

Fadi P. Deek and James A. M. McHugh (2008). *Open Source: Technology and Policy*. Cambridge University Press.

Joshua Gay, ed. (2002). *Free Software, Free Society: Selected Essays of Richard M. Stallman*. GNU Press, Free Software Foundation.

8

E-Government: Opportunities and Challenges

The initiatives of government agencies and departments to use ICT tools and applications, Internet and mobile devices to support good governance, strengthen existing relationships and build new partnerships within civil society, are known as eGovernment initiatives. As with e-commerce, eGovernment represents the introduction of a great wave of technological innovation as well as government reinvention. It represents a tremendous impetus to move forward in the 21st century with higher quality, cost effective government services and a better relationship between citizens and government. Many government agencies in developed countries have taken progressive steps toward the web and ICT use, adding coherence to all local activities on the Internet, widening local access and skills, opening up interactive services for local debates, and increasing the participation of citizens on promotion and management of the territory.

The potential for eGovernment in developing countries, however, remains largely unexploited, even though. ICT is believed to offer considerable potential for the sustainable development of eGovernment. Different human, organizational and technological factors, issues and problems pertain in these countries, requiring focused studies and appropriate approaches. ICT, in general, is referred to as an "enabler", but on the other hand it should also be regarded as a challenge and a peril in itself. The organizations, public or private, which ignore the potential value and use of ICT may suffer pivotal competitive disadvantages. Nevertheless, some eGovernment initiatives have flourished in developing countries too, e.g.

Brazil, India, Chile, etc. What the experience in these countries shows, is that governments in the developing world can effectively exploit and appropriate the benefits of ICT, but eGovernment success entails the accommodation of certain unique conditions, needs and obstacles. The adaptive challenges of eGovernment go far beyond technology, they call for organizational structures and skills, new forms of leadership, transformation of public-private partnerships.

The advent of the Internet, digital connectivity, the explosion and use of e-commerce and e-business models in the private sector are pressuring the public sector to rethink hierarchical, bureaucratic organizational models. Customers, citizens and businesses are faced every day with new innovative e-business and e-commerce models implemented by the private sector and made possible by ICT tools and applications, are requiring the same from governmental organizations.

Osborne and Gaebler referred to citizens as customers for governments, since governments need to empower rather than serve, to shift from hierarchy to teamwork and participation, to be mission oriented and customer focused, and to focus on prevention rather than cure. Governments worldwide are faced with the challenge of transformation and the need to modernize administrative practices and management systems. Recently, the public sector has began to recognize the potential opportunities offered by ICT and e-business models to fit with citizens' demands, to offer better services to citizens and to increase efficiency by streamlining internal processes.

Tapscott and Caston argue that ICT causes a "paradigm shift" introducing "the age of network intelligence", reinventing businesses, governments and individuals. Paradigm shifts prevail in the public sector too. The traditional bureaucratic paradigm, characterized by internal productive efficiency, functional rationality, departmentalization, hierarchical control and rule-based management, is being replaced by competitive, knowledge based economy requirements, such as: flexibility, network organization, vertical/horizontal integration, innovative entrepreneurship, organization learning, speed up in service delivery, and a customer driven strategy.

These new paradigms thrust the shift toward eGovernment paradigm, which emphasizes coordinated network building, external collaboration and customer services.

EGovernment means different things for different people. Some simply define it as digital governmental information or a way of engaging in digital transactions with customers. For others eGovernment simply consists of the creation of a web site where information about political and governmental issues is presented. These narrow ways of defining and conceptualizing eGovernment restrict the range of opportunities it offers. One of the reasons why many eGovernment initiatives fail is related to the narrow definition and poor understanding of the eGovernment concept, processes and functions. EGovernment is a multidimensional and complex concept, which requires a broad definition and understanding, in order to be able to design and implement a successful strategy.

The crucial element of all these definitions is the use of ICT tools to reinvent the public sector by transforming its internal and external way of doing things and its interrelationships with customers and the business community. The analysis of these definitions allows us to individuate the main issues and components that characterize an eGovernment framework, such as:

1. Transformation areas (internal, external, relational);
2. Users, customers, actors and their interrelationships;
3. eGovernment application domains.

Transformation Areas

The above definitions encompass three critical transformation areas of eGovernment:

1. *Internal* - which refers to the use of ICT to improve the efficiency and effectiveness of internal functions and processes of government by interrelating different departments and agencies. Thus, information can flow much faster and more easily among different governmental departments, reducing processing time, paperwork bottlenecks, and eliminating long, bureaucratic and inefficient approval procedures. Internetworking among different go vernmental departments improves internal efficiency by enabling time reductions for using, storing and collecting data, reduction of labor costs and information handling costs, as well as the speed and accuracy of task processing.
2. *External* - ICT opens up new possibilities for governments to be more transparent to citizens and businesses, giving access to a greater range

of information collected and generated by government. ICT creates also opportunities for partnership and collaboration among different governmental institutions. Electronic government blurs the lines not only within government agencies, but also between government and those that touch it.

3. *Relational* - ICT adoption may enable fundamental changes in the relationships between the citizens and the state, and between nation states, with implications for the democratic process and structures of government. Vertical and horizontal integration of services can be realized, enabling the integration of information and services from various government agencies to help citizens and other stakeholders get seamless services. Fountain uses the concept of the "virtual state" that is a governmental entity organized with "virtual agencies, cross agencies, public- priva te networks whose structures and capacity depend on the Internet and web".

According to these three transformational areas it is obvious that an eGovernment initiative does not consist of a simple business process reengineering. Indeed, as Tapscott suggests: "It requires a radical rethinking of the nature and functioning of the organization and the relationships between organizations. It needs to focus in a web of relationships including all levels and business functions, in which the boundaries inside and outside are permeable and fluid".

EGovernment Web of Interrelationships

The target of eGovernment encompasses four main groups: citizens, businesses, governments (other governments and public agencies) and employees. The electronic transactio ns and interactions between government and each group constitute the eGovernment web of relationships and the respective four main blocks of eGovernment, that are:

1. Government to Citizens (G2C)
2. Government to Business (G2B)
3. Government to Government (G2G)
4. Government to Employees (G2E)

Most researchers and academics refer only to the first three blocks, without considering the fourth or simply including it as part of 'government to government' block. The relationships, interactions and transactions between government and employees in fact constitute another large eGovernment block, which requires a separate and very careful handling. Many people today refer to employees as internal customers and as a result, in order for an eGovernment initiative to be customer oriented and centric, it has to take into account needs and requirements of this group as well. More specifically, these eGovernment blocks can be characterised as follows:

1. *Government to Citizen*: deals with the relationship between government and citizens. EGovernment allows government agencies to talk, listen, relate and continuously communicate with its citizens, supporting, in this way, accountability, democracy and improvements to public services. A broad array of interactions can be developed ranging from the delivery of services and the provision of welfare and health benefits to regulatory and compliance oriented licensing. G2C allows customers to access government information and services instantly, conveniently, from everywhere, by use of multiple channels (PC, Web TV, mobile phone or wireless device). It also enables and reinforces their participation in local community life (send an email or contribute to an online discussion forum).
2. *Government to Business:* consists of the electronic interactions between government agencies and private businesses. It allows e-transaction initiatives such as e- procurement and the development of an electronic marketplace for government. Companies everywhere are conducting business-to-business e-commerce in order to lower their costs and improve inventory control. The opportunity to conduct online transactions with government reduces red tape and simplifies regulatory processes, therefore helping businesses to become more competitive. The delivery of integrated, single-source public services creates opportunities for businesses and government to partner together for establishing a web presence faster and cheaper.
3. *Government to Government:* refers to the relationship between governmental organizations, as for example national, regional and local governmental organizations, or with other foreign government organizations. Governments depend on other levels of government within the state to effectively deliver services and allocate respons ibilities. In order to realize a single access point, collaboration and cooperation among different governmental departments and agencies is compulsory. Online communication and cooperation allows

government agencies and departments to share databases, resources, pool skills and capabilities, enhancing the efficiency and effectivity of processes.

4. *Government to Employees*: refers to the relationship between government and its employees. G2E is an effective way to provide e-learning, bring employees together and to promote knowledge sharing among them. It gives employees the possibility of accessing relevant information regarding: compensation and benefit policies, training and learning opportunities, civil rights laws, etc. G2E refers also to strategic and tactical mechanisms for encouraging the implementation of government goals and programs as well as human resource management, budgeting and accounting. The full exploitation and implementation of these complex webs of inter-relationships requires three main application domains for eGovernment:
   - e-Administration – for automation and computerization of administrative tasks and for realization of strategic connections among internal processes, departments and functions. e-Citizens and e-Services – to realize connections and interrelationships among governments and citizens and to deliver automated services.
   - e-Society – to enable relationships and interactions beyond boundaries, among public agencies, private sector and civil community in general.

These three application domains should be considered as overlapping and eGovernment can be found in the overlapping area of these three application domains, demonstrating the complexities and heterogeneities needed to be handled for assuring its success.

It is now widely accepted that ICT offers increased opportunities for economic development and plays a critical role in rapid economic change, productive capacity improvements and international competitiveness enhancement for developing countries. The range of choices and opportunities in developing countries is expanding. ICT is believed to be a powerful enabling tool to address some of the key barriers and challenges for entering the global economy and for future growth potential. It can transform old challenges and create unprecedented possibilities for sustainable economic development, just as it has done for businesses in the industrial world. ICTs offer the potential not just to collect, store, process and diffuse enormous quantities of information at minimal cost, but also to network, interact and communicate across the world. Econometric studies have found evidence of a strong positive relationship between ICT investments and GDP growth illustrating the importance of ICTs for development, both in the commercial and the public sectors. An OECD research project, based on national studies about the impact of ICT on the economy, has shown that ICT investments accounted for between 0.5% and 1.3% in GDP growth per capita per annum over a number of economies in the 1995–2000 period. In Australia, the growth was 1.3% per annum over 1996-2000. More detailed statistics are set out in Table 2.

Nevertheless, ICT represents a high risk for developing countries – the risk to deepen the digital divide and to further marginalize them with the networking revolution. Countries which fail to embrace and use ICT tools for entering the global network and for addressing developme nt needs, will suffer pivotal disadvantages in form of information poverty that could further widen the gap in economic status and competitiveness.

Recognizing the power of ICTs, many developing countries, assisted by international organizations for development, have started building and encouraging e-strategies and initiatives to address a wide range of economic, social, technological, infrastructural, legal and educational issues. G8 on Digital Opportunities Task Force and UN Task Force on ICT Access have evidenced eGovernment as one of the priorities based on the decisive role it would play for ICT accessibility. Consequently, eGovernment initiatives have flourished in many developing countries such as Brazil, India, Chile, Argentina, the Philippines, etc., reaping the advantages and opportunities of ICT and the knowledge economy for collaboration, networking, better services, efficiency and effectivity. However, many other initiatives have failed. In a recent survey regarding the success and failure rates of eGovernment in developing and transitional countries, Heeks found that more than one-third of initiatives are total failures (e.g. the failure of decision support systems in East Africa); further, half can be cons idered to be partial failures (e.g. the partial failure of management information systems in Eastern Europe); and roughly one-seventh are successes.

An eGovernment benchmark study conducted by the American Society for Public Administration aimed to categorize the progress made by developed and developing countries in developing an online presence on a five stages scale: Emerging stage; Enhanced stage; Interactive stage; Transactional stage and Seamless stage, revealed that nearly all 32 countries at the Emerging Presence level were among the world's least developed nations, characterized by static and insufficient information that is infrequently updated, few interactive features, and non-existent online services. There were, however, several cases of developing countries that were at the Enhanced Presence stage and Interactive Presence stage, demonstrating their predisposition for eGovernment success. The Transactional stage and in particular the Seamless stage were seen as very distant points for both developed and developing countries, with a few exceptions. The prevalence of stages 1 and 2 among developing countries can be explained by the fact that the adaptive challenges of eGovernment go far beyond technology: they call for organizational structures and skills, new leadership forms, and the transformation of public-private relationships.

Based on these assertions, in this paper I have tried to evaluate and assess the potential of eGovernment projects and initiatives for developing countries as well as to identify and analyze the principle challenges that may be encountered. An empirical web-based study of 15 eGovernment case studies in 9 developing countries was performed and the results obtained are presented in the following sections. What the experience with eGovernment initiatives in developing countries shows, is that:

EGovernment offers great potential and opportunity for developing countries for improving their governance and citizen satisfaction level. However, in order to realize the full potential of these applications, governments must accommodate certain unique conditions and be able to manage a set of issues, problems and related challenges.

Opportunities of EGovernment

Benefits assured by use and application of eGovernment in developing countries are the same as those in developed countries. The differences between these two groups could result from the fact that many potential benefits of eGovernment are not reaped by developing countries as consequence of their limited use of eGovernment.

Table 1 shows the main opportunities eGovernment has triggered in our analyzed cases.

Table 1. EGovernment Opportunities

1. Cost reduction and efficiency gains
2. Quality of service delivery to businesses and customers
3. Transparency, anticorruption, accountability
4. Increase the capacity of government
5. Network and community creation
6. Improve the quality of decision making
7. Promote use of ICT in other sectors of the society

Cost Reduction and Efficiency Gains

Researchers agree that ICT has considerable potential to contribute to efficiency gains and cost reductions for private organizations. Furthermore, these benefits constitute a major aspect of eGovernment initiatives. Putting services on- line substantially decreases the processing costs of many activities compared with the manual way of handling operations. For example, it costs the US Inland Revenue Service $1.60 to process a paper tax form, but only $0.40 to process an electronic form. The appropriate application of ICT may possibly reduce the number of inefficiencies in processes by allowing file and data sharing across government departments, thereby contributing to the elimination of mistakes from manual procedures, reducing the required time for transactions. Efficiency is also attained by streamlining internal processes, by enabling faster and more informed decision making, and by speeding up transaction processing.

Quality of Service Delivery to Businesses and Customers

In the traditional model of public service delivery, the procedures are long, time consuming and lack transparency. A business that wishes to obtain a license or a permit has to fill out a number of application forms, has to visit a number of different offices and spend a considerable amount of time. If a citizen wishes to be issued with a certificate or any other official document, he or she will have to travel to the central government office, go to different offices and spend a lot of time for a simple service. The consequences are high costs and citizen and business dissatisfaction. An eGovernment initiative, on the other hand, which puts government services online, thereby reducing the bureaucracy, offers roudn the clock accessibility, fast and convenient transactions, and obviously enhances the quality of services, in terms of time, content and accessibility.

Example: In Bahia, Brazil, Citizen Assistance Service Centres have been created offering over 500 separate services. These centres are placed in shopping malls or other public places, and people going to shop can simultaneously apply for different public services such as getting an identification card, looking for a new job, getting a passport, and checking on their retirement eligibility. Customer satisfaction studies revealed that over 89% of citizens evaluated the service centers as excellent. Thus, the quality of services is ensured by the reduced time that users spend on getting official documents, waiting and queuing to get documents, traveling, as well as more customized products and services, error free documents, and 24 × 7 × 365 accessibility.

Transparency, Anticorruption and Accountability

EGovernment helps to increase the transparency of decision-making processes. In many cases eGovernment offers opportunities for citizens to directly participate in decision- making, by allowing them to provide their own ideas and suggestions in forums and on- line communities. If web sites are designed carefully and openly, they can be valuable resources for transparency as citizens, businesses and other stakeholders should be able to see political and governmental information, rules and policies. Previously it was often necessary to go directly to governmental offices to obtain information, but now this information should be available on the web. The availability of a diversity of publications regarding the activities of the public administration, as well as economic and legislative aspects, increases the transparency too.

Increase the Capacity of Government

The use of ICT for the reorganization of internal administration transactions, communications, interrelationships and for easy information flow and transfer offers considerable opportunity to increase government capacity. Intranets allow different departments to share databases of common customers and to pool skills and capacities of their members for problem solving. These facilities in turn will pledge faster information flow and transfer, quicker and cheaper provision of goods and services, faster and better decision making processes, and unplugged paper bottlenecks. Knowledge based or expert systems help to create a more responsive and guideline based process. This approach assures benefits for businesses, which become both consumers of government services and providers of goods and services to the government. It also assures benefits to the government itself through reduced costs and spending, which could require lower taxes to finance.

Example: The Time Saver Centre in Sao Paulo, Brazil, brings together multiple services in a single location. Its objective is to deliver services more quickly and to increase the satisfaction level of its citizens. A person requiring a service, on reaching the appropriate agency, can register in the computerized tracking system and receive an electronic ticket, which indicates the services desired and the estimated waiting time. They can receive at the same time different services that traditionally were separated such as vehicle registration, driver's license, identification card, unemployment insurance etc. A customer satisfaction survey conducted in 2000 for five centres reveals that 94% of respondents evaluate services as "excellent" or "good". This case demonstrates the remarkable improvements that can be realized in service delivery.

Network and Community Creation

ICT creates both pressures and opportunities for network creation and community building. As argued before, an eGove rnment initiative requires a complex web of interrelationships among government, customers, businesses, employees and other governmental agencies. Moreover, the very nature and function of eGovernment require a network approach to put together skills, technologies, information and knowledge that span the boundaries of different governmental agencies. It is generally impossible to find all of them in one single governmental agency. The need for learning and training, for example, requires a partnership between government agencies/departments and universities or research institutions. The provision of integrated services at one contact point requires the cooperation and collaboration of different departments and agencies, horizontal and vertical integration, and therefore the creation of a large and diversified network of relationships. The successful use and diffusion of ICTs in the public sector involves a collective, multidisciplinary and dynamic learning process. Moreover, the realization of electronic transactions triggers network creation among private companies, financial institutions, telecommunication and ISPs. On the other hand, an eGovernment initiative enables community creation, giving citizens and businesses the possibility to participate in forums, and in decision making processes, contributing actively to different political and governmental discussions.

Example: Columbia's government portal is the entry point to every government agency website in the country, allowing citizens to search for and consult government information and to e- mail government representatives either to complain about problems or to make suggestions. A specific unit, the Government online Network, composed of eight people trained in the technology of government portals, was created for realizing Columbia's website and for advising, supporting, training and monitoring the remainder federal government. Financial support was provided by the UNDP, while the technology and experience were provided primarily by a partnership with two private companies: GovWorks Latin America/Taillon and Arthur Andersen.

Improve the Quality of Decision Making

Community creation, forums, continuous interaction and communication between government and its citizens contribute further to the decision making process. By means of active participation in political and government discussions, citizens can contribute their own ideas, and share their knowledge and information. This will in turn lead to building trust in government and improving the relationships between the government and the governed. The OECD argues that the strengthening relationship between government and citizens could improve the quality of services by allowing government to tap wider sources of information, perspectives and solutions to meet the challenges of policy making under conditions of increased complexity. Considering citizens as governmental customers, listening and understanding to their needs and requirements, is essential for a better decision-making process. The appropriate use of shared data and information by all governmental agencies and departments offers the possibility to make quick decisions thus to serve the community better. However improvements in the speed and quality of decision making depend greatly on the willingness of governments to be empowered with new information, the capability of staff to process the large amount of information, the prevailing cultural values as well as the motivation of governments to shift from a hierarchical public administration model to a flexible, less centralized model.

Example: The CRISTAL initiative of Argentina's government was launched in order to disseminate information regarding the use of public funds, including information about the amounts of money for different programs, financial and employment data, public debt account including terms, guarantees, interest costs, and the outstanding tax and customs obligations of private companies. Its primary goal is to inform customers/ citizens, to disseminate content and information, empowering customers to exercise more control over their political representatives. In their web site they also provided a specific section where users can send their questions, comments and suggestions for further improvements. Their feedback allows the government to adjust the content and information, to customize the information and to reorganize itself around customers' needs and requirements.

Promote Use of ICT in Other Sectors of the Society

Continuous interaction and communication between government and its stakeholders contributes to the creation of awareness about the potential contribution of ICT to local community activities. In this way, eGovernment plays a vital role, not only in facilitating market- led initiatives but also in initiating the process of capability building and in coordinating the actions of a large number of interested stakeholders. In fact, one of the main benefits of an eGovernment initiative consists of the promotion of ICT use in other sectors. In order for eGovernment staff to interact, transact and communicate electronically with businesses, citizens and other stakeholders, it is necessary to mandate the use of ICT tools and applications. For a government-to-business electronic transaction to occur, the business itself needs to make use of electronic equipment. On the other hand, financial institutions have to create secure and reliable methods for electronic transactions. The development of new technological and management capacities required for eGovernment functionality encourage the development in turn of new training courses and modules in schools and universities trying to supply the required skills and capabilities to the market.

Example: In India, the Gyandoot project is a government-to-citizen intranet project which offers numerous benefits to the region, to citizens and to the community in general. The goal of the project has been to establish community owned technologically innovative and sustainable information kiosks in a poverty-stricken rural area of Madhya Pradesh. The benefits assured by this intranet system have increased the awareness of ICT importance and have spin off other IT initiatives and programs, such as: the creation of new private ICT training institutions; a high level of student enrolment – about 60%; parliament has allocated resources to set up other kiosks in schools and to develop new models for e-education; Indira Gandhi National Open University has opened a study center for undergraduate and postgraduate courses on computer applications; the government has instituted a cash award to motivate ICT projects.

Challenges for a Implementing of eGovernment Initiatives

While it is evident that eGovernment and ICTs, in general, are powerful drivers of wealth creation and growth, there remain many challenges which hamper the exploration and exploitation of its opportunities. The multidimensionality and complexity of eGovernment initiatives implies the existence of a wide variety of challenges and barriers to its implementation and management. Table 2 represents the main challenges, identified in the case study analysis for eGovernment development and impleme ntation in developing countries. In this section, we briefly introduce each of these challenges and offer some policy implications for their management.

Table 2. EGovernment Challenges

1.	ICT infrastructure (e-readiness, computer literacy, telecommunication equipment)
2.	Policy issues (legislation)
3.	Human capital development and life long learning (skills, capabilities, education, learning)
4.	Change management (culture, resistance to change)
5.	Partnership and collaboration (public/private partnership, community and network creation)
6.	Strategy (vision, mission)
7.	Leadership role (motivate, involve, influence, support)

ICT Infrastructure

ICT infrastructure is recognised to be one of the main challenges for eGovernment. Internetworking is required to enable appropriate sharing of information and open up new channels for communication and delivery of new services. For a transition to electronic government, an architecture, that is, a guiding set of principles, models and standards, is needed. Many developing countries suffer from the digital divide, and they are not able to deploy the appropriate ICT infrastructure for eGovernment deployment. The digital divide between richer countries and developing ones is large with high- income economies having 416 personal computers per 1,000 people and low-income economies only 6 per 1,000. The development of basic infrastructure to capture the advantages of new technologies and communications tools is essential for implementing eGovernment. Different access methods, such as remote access by cellular phones, satellite receivers, kiosks, etc., need to be taken into consideration by governments in order that all members of society can be served irrespective of their physical and financial capabilities.

However, an ICT infrastructure does not consist simply of telecommunications and computer equipment. E-readiness and ICT literacy are also necessary in order for people to be able to use and benefit from eGovernment applications. Having the education, freedom and desire to access information is critical to eGovernment efficacy. Presumably, the higher the level of human development, the more likely citizens will be inclined to accept and use eGovernment services.

*Example: I*n the Gyandoot project, the poor infrastructure facilities constituted one of the major problems encountered in developing and implementing the project. Local rural telephones infrastructure did not operate with optical fiber cable, and in consequence there were initially significant reliability problems. This caused a decrease in the motivational level of kiosk managers to participate in the project. Substantial problems were encountered with literacy and skills to use new technological tools and applications. To ensure the success and the sustainability of the project, the Indian telecommunications department undertook actions to upgrade the level and quality of connections, as well as study alternative solutions (such as wireless applications) to cover those zones where telephones were not available. In addition some basic training was provided to people who were directly engaged in management and maintenance of kiosks.

Policy Issues

Processing of eGovernment principles and functions requires a range of new rules, policies, laws and legislative changes to address electronic activities including electronic signatures, electronic archiving, freedom of information, data protection, computer crime, intellectual property rights and copyright issues. Dealing with eGovernment means signing a contract or a digital agreement, which has to be protected and recognized by a formalized law, which protect and secure these kinds of activities or processes. In many developing countries, e-business and eGovernment laws are not yet available.

Establishing protections and legal reforms will be needed to ensure, among other things, the privacy, security and legal recognition of electronic interactions and electronic signatures. Hence, governments all over the world need to tackle the design and development of a public key infrastructure, which will guarantee secure transactions between organizations and individuals.

Example: In the e-procurement system initiative in the Philippines, which aimed to streamline the purchase of goods and services for a large number of government departments and agencies, a number of actions were undertaken to change the legal framework and to issue new rules and policies that govern and regulate electronic commerce and interactions. An executive order was issued which provided legal guidelines about how to conduct electronic business, and how to advertise and post bids or notices in the new electronic system. In addition, an e-Commerce law was promulgated, in order to give legal protection to electronic documents.

Human Capital Development and Life Long Learning

A major challenge of an eGovernment initiative is the lack of ICT skills in the public sector. This is a particular problem in developing countries, where the chronic lack of qualified staff and inadequate human resources training has been a problem for years. The availability of appropriate skills is central for successful eGovernment implementation. EGovernment requires hybrid human capacities: technological, commercial and management. Technical skills for installation, maintenance, designing and implementation of ICT infrastructure, as well as skills for using and managing online processes, functions and customers, are necessary. To address human capital development issues, knowledge management initiatives are required focusing on staff training, seminars, workshops in order to create the basic skills for eGovernment handling.

Example: In Beijing's Business e-Park initiative, a key step in project implementation was the education programme. Firstly government officers learnt to do their jobs more quickly and efficiently. It wa s also important to educate government leaders, as they were responsible to explain what eGovernment is and what its benefits will be for the community. Ultimately, basic computer and Internet training were provided to government staff and public users of the eGovernment system.

In general, in almost all cases the focus on training and education programs was a paramount phase for the assurance of project endurance. However, the human capability development doesn't end up with the acquisition and achievement of basic initial skills. Instead, lifelong learning is an essential prerequisite as the rate of change increases and new technologies, practices and competitive models emerge. The full economic benefits of IT depend on a process of social experimentation and learning, which is still at an early stage.

Change Management

Change management issues must be addressed as new work practices, new ways of processing and performing tasks are introduced. EGovernment correctly designed doesn't simply save costs and improve service quality; instead it revolutionizes and reinvents the government processes and functions. Change management can be divided into two sub-concepts: Change Management Approach and Management of Resistance to Change. Change management approach refers to the change management procedures established within organizations. DeLisi identifies culture as the primary driver of strategic organizational change. Being aware of an organization's culture is already a big step towards a higher capacity to change. Hierarchy is the most traditional of cultural values of a government bureaucracy, in many ways its defining feature. In particular, intranets and the sharing of information throughout organizations can challenge hierarchies and can only really benefit an organization that develops a more networked approach; ICT is distinguished by its network character.

Employee resistance to change is still the biggest barrier to successful change. Employees fear changes in general and ICT applications in particular as they believe that ICT would replace them and so cause job losses. Moreover, it is very difficult in a short time to turn off traditional methods of working and learn new ones. Addressing resistance successfully means ensuring the existence of incentives for employees to learn and change and the establishment of well-structured plans that embrace employee participation throughout all stages of a change process.

Example: It is relevant to mention here the experience in India of the Vijaywada Online Information Center (VOICE). The main objective of this initiative was to realize an electronic system which enables the delivery of municipal services such as buildingapprovals, status certificates, and handles the collection of different types of taxes. Resistance to change from public staff was one of the major problems encountered in this endeavor. The revenue department staff were those who caused most problems as they stood to lose the income received from bribes. Some staff feared job losses, some others were reluctant to learn and use the new technology and new work practices. Several meetings and performance reviews were organized to persuade staff to become accountable and to motivate them for better performance.

Partnership and Collaboration

Collaboration and cooperation at local, regional and national levels, as well as between public and private organizations, are important elements in the eGovernment development process. Nevertheless, collaboration and cooperation are not simple to realize. Governments often exhibit considerable resistance to open and transparent systems as they try to preserve their authority, power and hierarchical status. Citizens distrust their governments, especially where there has been a history of dictatorship, political instability or large-scale corruption. To ensure that the public and stakeholders will be partners in the eGovernment effort, it is important to try to build trust in government. Collaboration between the private (assuming that there is a private sector) and public sectors is needed too, in order to provide resources, skills and capabilities that the government lacks. For example, the ICT private sector is able to support government with technical skills and infrastructure; meanwhile, universities will provide the required staff, learning and training courses for government staff and citizens, and other governmental departments and agencies can contribute in data and information flow and knowledge sharing for problem solving of similar tasks or processes and so on. A 'new' development model is emerging that focuses on partnership among stakeholders in the knowledge-based development program.

Example: The initiative of the State of Andhra Pradesh in India to computerize the 1,124 administrative units, called mandals, in order to realize online delivery of services, required strong coordination and collaboration between various departments. Different databases were handled and managed by different departments, one from the revenue department, one from the national informatics department, another from the social welfare department. These departments were geographically spread over an area of 275,000 sq. kms., but the timeless delivery of services required the instant collaboration, communication and interaction between them.

Strategy

One of the main challenges for an eGovernment project is the establishment of an appropriate and context tailored strategy. Every project or initiative needs to be rooted in a very careful, analytical and dynamic strategy. This seems to be a very difficult task, requiring a focus on many aspects and processes, a holistic vision, long-term focus and objectives. Many public institutions limit their activities to a simple transfer of their information and services online without taking into consideration the re-engineering process needed to grasp the full benefits.

The government must have a clear strategy to overcome the barriers to change. Part of the strategy is to engage in a rigorous assessment of the current situation, the reality on the ground and the inventory of projects, articulate costs, impacts and benefits of programme as well as continuously monitor and evaluate the project upgrading. Borrowing a lesson from the private sector, eGovernment must be customer-driven and service oriented, meeting the needs of citizens and improving the quality of life. This means that a vision of eGovernment implies providing greater access to information as well as better, more equal services and procedures for public and businesses.

Even when eGovernment projects seek to improve internal government processes, the end goal should be making government serve citizens better. This means recognizing the diverse roles that citizens can play as partners, taxpayers, constituents, employers, employees, students, investors and lobbyists.

Example: A critical point for the initiative of the government of Colombia to realize an eGovernment portal was the definition of strategy and actions to be pursued. Initially the government created a specific unit to develop the strategy, objectives, and plan for actions and afterwards to assist and monitor the work progress. A 'Connectivity Agenda' was formulated which specifies the key objectives of the initiative, the strategic framework for subsequent actions, and different projects to undertake. The Agenda established an action framework that guided any plan in Colombia related to ICT development and electronic applications, thus allowing a rational and coordinated investment effort.

Leadership Role

The public sector presents unique challenges for leadership. Changing and hazy visions confuse expectations for reforms and leaders. Leadership is one of the main driving forces of every new and innovative project or initiative. Since eGovernment is a complex process, accompanied by high costs, risks and challenges, public organizations are generally resistant to the initiation of change. A leading player (organization, institution), which is able to understand the real costs and benefits of the project, to motivate, influence, include and support other organizations and institutions, is required. Leadership is necessary before, during and after project implementation. Before the project is initiated, leadership is needed in order to explain the concept, the model and create awareness; during the project, leadership is needed to manage change and support the project; and after the project, it is needed to pledge the required flexibility and adaptability of the initiative. Top leadership involvement and clear lines of accountability for making management improvements are critical to overcoming organizations' natural resistance to change, marshalling the resources needed to improve management, and building and maintaining the organization wide commitment to new ways of doing government.

Example: Chile's government procurement e-System initiative was seen as a technocratic solution. As a consequence, the pledge of strong political support and top leadership was a critical issue. The political support through exposure in the press outlining the benefits of the initiative in terms of transparency, efficiency and e-commerce capacity was sought as vital by organizing staff. Further leadership and support was provided by lobbying political parties, interest groups, private sector advocates and information technology companies.

Internet Activism

Internet activism is the use of electronic communication technologies such as e-mail, the World Wide Web, and podcasts for various forms of activism to enable faster communications by citizen movements and the delivery of local information to a large audience. Internet technologies are used for cause-related fundraising, community building, lobbying, and organizing.

The Internet is a key resource for independent activists or E-activists, particularly those whose message may run counter to the mainstream. "Especially when a serious violation of human rights occurs, the Internet is essential in reporting the atrocity to the outside world," Listservs like BurmaNet help distribute news that would otherwise be inaccessible in these countries.

Internet activists also pass on E-petitions to be sent to the government and public and private organizations to protest against and urge for positive policy change in areas from the arms trade to animal testing. Many non-profits and charities use these methods, emailing petitions to those on their email list, asking people to pass them on. The Internet also enables organizations such as NGOs to communicate with individuals in an inexpensive and timely manner. Gatherings and protests can be organized with the input of the organizers and the participants. Lobbying is also made easier via the Internet, thanks to mass e-mail and its ability to broadcast a message widely at little cost. Vegh's concept of organization/mobilization, for example, can refer to activities taking place solely online, solely offline but organized online, or a combination of online and offline. Mainstream social-networking sites, most noticeably Facebook.com, are also making e-activist tools available to their users.

In addition, Denial-of-service attacks, the taking over and vandalizing of a website, uploading Trojan horses, and sending out an e-mail bomb (mass e-mailings) are also examples of Internet activism.

Hacktivism One of the earliest know uses of the Internet as a medium for activism was that around Lotus MarketPlace. On April 10, 1990, Lotus announced a direct-mail marketing database product that was to contain name, address, and spending habit information on 120 million individual U.S. citizens. While much of the same data was already available, privacy advocates worried about the availability of this data within one database. Furthermore, the data would be on CD-ROM, and so would remain fixed until a new CD-ROM was issued.

In response, a mass e-mail and E-bulletin-board campaign was started, which included information on contacting Lotus and form letters. Larry Seiler, a New England-based computer professional posted a message that was widely reposted on newsgroups and via e-mail: "It will contain a LOT of personal information about YOU, which anyone in the country can access by just buying the discs. It seems to me (and to a lot of other people, too) that this will be a little too much like big brother, and it seems like a good idea to get out while there is still time."Over 30,000 people contacted Lotus and asked for their names to be removed from the database. On January 23, 1991, Lotus announced that it had cancelled MarketPlace.

In 1993 a survey article about online activism around the world, from Croatia to the United States appeared in *The Nation* magazine, with several activists being quoted about their projects and views. In 1995-1998, Z magazine offered courses online through Left Online University, with being taught on "Using the Internet for Electronic Activism."

Another well-known example of early Internet activism took place in 1998, when the Mexican rebel group EZLN used decentralized communications, such as cell phones, to network with developed world activists and help create the anti-globalization group Peoples Global Action (PGA) to protest the World Trade Organisation (WTO) in Geneva. The PGA continued to call for "global days of action" and rally support of other anti-globalization groups in this way. Later, a worldwide network of Internet activist sites, under the umbrella name of Indymedia, was created "for the purpose of providing grassroots coverage of the WTO protests in Seattle" in 1999. Dorothy Kidd quotes Sheri Herndon in a July 2001 telephone interview about the role of the Internet in the anti-WTO protests: "The timing was right, there was a space, the platform was created, the Internet was being used, we could bypass the corporate media, we were using open publishing, we were using multimedia platforms. So those hadn't been available, and then there was the beginning of the anti-globalization movement in the United States."

When discussing the 2004 U.S. presidential election candidates, Carol Darr, director of the Institute for Politics, Democracy & the Internet at George Washington University in Washington, D.C., said of the candidates which benefited from use of the Internet to attract supporters: "They are all charismatic, outspoken mavericks and insurgents. Given that the Internet is interactive and requires an affirmative action on the part of the users, as opposed to a passive response from TV users, it is not surprising that the candidate has to be someone people want to touch and interact with."

A more decentralized approach to campaigning arose, in contrast to a top-down, message-focused approach usually conducted in the mainstream. "The mantra has always been, 'Keep your message consistent. Keep your message consistent,'" said John Hlinko, who has participated in Internet campaigns for MoveOn.org and the electoral primary campaign of Wesley Clark. "That was all well and good in the past. Now it's a recipe for disaster... You can choose to have a Stalinist structure that's really doctrinaire and that's really opposed to grassroots. Or you can say, 'Go forth. Do what you're going to do.' As long as we're running in the same direction, it's much better to give some freedom."

"The Internet is tailor-made for a populist, insurgent movement," says Joe Trippi, who managed the Howard Dean campaign. In his campaign memoir, *The Revolution Will Not Be Televised,* Trippi notes that the Internet's roots in the open-source ARPAnet, its hacker culture, and its decentralized, scattered architecture make it difficult for big, establishment candidates, companies and media to gain control of it. And the establishment loathes what it can't control.

This independence is by design, and the Internet community values above almost anything the distance it has from the slow, homogenous stream of American commerce and culture. Progressive candidates and companies with forward-looking vision have an advantage on the Internet, too. Television is, by its nature, a nostalgic medium. Look at Ronald Reagan's campaign ads in the 1980s—they were masterpieces of nostalgia promising a return to America's past glory and prosperity. The Internet, on the other hand, is a forward-thinking and forward-moving medium, embracing change and pushing the envelope of technology and communication.

According to some observers, the Internet may have considerable potential to reach and engage opinion leaders who influence the thinking and behavior of others. According to the Institute for Politics, Democracy & the Internet, what they call "Online Political Citizens" (OPCs) are "seven times more likely than average citizens to serve as opinion leaders among their friends, relatives and colleagues...Normally, 10% of Americans qualify as Influentials. Our study found that 69% of Online Political Citizens are Influentials."

The Internet has also made it easier for small donors to play a meaningful role in financing political campaigns. Previously, small-donor fundraising was prohibitively expensive, as costs of printing and postage ate up most of the money raised. Groups like MoveOn, however, have found that they can raise large amounts of money from small donors at minimal cost, with credit card transaction fees constituting their biggest expense. "For the first time, you have a door into the political process that isn't marked 'big money,' " says Darr. "That changes everything.

Corporations are also using Internet activist techniques to increase support for their causes. According to Christopher Palmeri with

BusinessWeek Online, companies launch sites with the intent to positively influence their own public image, to provide negative pressure on competitors, to influence opinion within select groups, and to push for policy changes.

The clothing manufacturer, American Apparel is an example: The company hosts a website called Legalize LA that advocates immigration reform via blog, online advertising, links to news stories and educational materials. Protest groups have responded by posting YouTube videos and establishing a boycott website.

Another example ActivistCash.org, which "is committed to providing detailed and up-to-date information about the funding source of radical anti-consumer organizations and activists." The website is created by the Center for Consumer Freedom, a "coalition of restaurants, food companies, and consumers" many of which have themselves been the target of the organizations which the website targets. This method of information dissemination is labelled "astroturfing," as opposed to "grassroots activism," due to the funding for such movements being largely private. More recent examples include the right-wing FreedomWorks.org which organized the "Taxpayer March on Washington" on September 12, 2009 and the Coaltion to Protect Patients' Rights, which opposes universal health care in the U.S.

Critics argue that this form Internet activism faces the same challenges as other aspects of the digital divide, particularly the global digital divide. Some say it gives disproportionate representation to those with disproportionate access or technological ability.

Another concern, expressed by author and law professor Cass Sunstein, is that online political discussions lead to "cyberbalkanization"—discussions that lead to fragmentation and polarization rather than consensus, because the same medium that lets people access a large number of news sources also lets them pinpoint the ones they agree with and ignore the rest.

The experience of the echo chamber is easier to create with a computer than with many of the forms of political interaction that preceded it," Sunstein told the *New York Times*. "The discussion will be about strategy, or horse-race issues or how bad the other candidates are, and it will seem like debate. It's not like this should be censored, but it can increase acrimony, increase extremism and make mutual understanding more difficult.

On the other hand, "the Internet connects all sides of issues, not just an ideologically broad anti-war constituency, from the leftists of ANSWER to the pressed-for-time 'soccer moms' who might prefer MoveOn, and conservative activists as well," opined Scott Duke Harris of the *San Jose Mercury News.* According to University of California professor Barbara Epstein, however, the Internet "allows people who agree with each other to talk to each other and gives them the impression of being part of a much larger network than is necessarily the case." She warns that the impersonal nature of communication by computer may actually undermine important human contact that always has been crucial to social movements.

Famed activist Ralph Nader has stated that "the Internet doesn't do a very good job of motivating action" citing that Congress, corporations and the Pentagon don't necessarily "fear the civic use of the Internet."

Hacktivism

Hacktivism is "the nonviolent use of illegal or legally ambiguous digital tools in pursuit of political ends. These tools include web site defacements, redirects, denial-of-service attacks, information theft, web site parodies, virtual sit-ins, virtual sabotage, and software development." It is often understood as the writing of code to promote political ideology—promoting expressive politics, free speech, human rights, or information ethics. Acts of hacktivism are carried out in the belief that proper use of code will be able to produce similar results to those produced by regular activism or civil disobedience.Hacktivist activities span many political ideals and issues. Freenet is a prime example of translating political thought (anyone should be able to speak) into code. Hacktivismo is an offshoot of Cult of the Dead Cow; its beliefs include access to information as a basic human right. The loose network of programmers, artists and radical militants 1984 network liberty alliance is more concerned with issues of free speech, surveillance and privacy in an era of increased technological surveillance.

Hacktivism is a controversial term, and can often be misconstrued as cyberterrorism. What separates hacktivism from cyberterrorism is a distinctly political or social cause behind the "haction". Some argue it was coined to describe how electronic direct action might work toward social change by combining programming skills with critical thinking. Others use it as practically synonymous with malicious, destructive acts that undermine the security of the Internet as a technical, economic, and political platform.

Essentially, the controversy reflects two divergent philosophical strands within the hacktivist movement. One strand thinks that malicious cyber-attacks are an acceptable form of direct action. The other strand thinks that all protest should be peaceful, refraining from destruction.

Some people describing themselves as hacktivists have taken to defacing websites for political reasons, such as attacking and defacing government websites as well as web sites of groups who oppose their ideology. Others, such as Oxblood Ruffin (the "foreign affairs minister" of Hacktivismo), have argued forcefully against definitions of hacktivism that include web defacements or denial-of-service attacks.

Critics suggest that DoS attacks are an attack on free speech; that they have unintended consequences; that they waste resources; and that they could lead to a "DoS war" which nobody will win. In 2006, Blue Security attempted to automate a DoS attack against spammers; this led to a massive DoS attack against Blue Security which knocked them, their old ISP and their DNS provider off the internet, destroying their business.

Depending on who is using the term, hacktivism can be a politically constructive form of anarchist civil disobedience or an undefined anti-systemical gesture; it can signal anticapitalist or political protest; it can denote anti-spam activists, security experts, or open source advocates. Critics of hacktivism fear that the lack of a clear agenda makes it a politically immature gesture, while those given to conspiracy theory hope to see in hacktivism an attempt to precipitate a crisis situation online.

Elements of Hacktivist "Hactions"

A Haction usually has the following elements.

1) Politically motivated
2) Place a premium on humor, and often resembles a digital form of clowning
3) Owns a moderate "Outlaw Orientation" as opposed to severe
4) The result of aggressive policy circumvention—rather than a gradual attempt to change a policy
5) Always non-violent- a haction never places another in direct danger
6) Capacity for solo activity—while most forms of political activism require the strength of masses, hacktivism is most often the result of the power of one, or small group.

7) Is most often carried out anonymously, and can take place over transnational boarders.

Forms of Hacktivism

In order to carry out their operations, hacktivists use a variety of software tools readily availble on the internet. In many cases the software can be downloaded from a popular website, or launched from a website with click of a button. Some of the more well known hacktivist tools are below:

1. Defacing Web Pages Between 1995-1999 Attrition.org reported 5,000 website defacements. In such a scenario, the hacktivist will significantly alter the front page of a company's or governemental agency's website.
2. Web Sit-ins In this form of hacktivism,hackers attempt to send so much traffic to the site, that the overwhelmed site becomes inaccessible to other users.
3. E-mail Bombing Hacktivists send scores of e-mails with large file attachments to their targets e-mail address.

Internet Security and Censorship Issues

Internet censorship is control or suppression of the publishing or accessing of information on the Internet. The legal issues are similar to offline censorship.

One difference is that national borders are more permeable online: residents of a country that bans certain information can find it on websites hosted outside the country. A government can try to prevent its citizens from viewing these even if it has no control over the websites themselves.

Filtering can be based on a blacklist or be dynamic. In the case of a blacklist, that list is usually not published. The list may be produced manually or automatically.

Barring total control on Internet-connected computers, such as in North Korea, total censorship of information on the Internet is very difficult (or impossible) to achieve due to the underlying distributed technology of the Internet. Pseudonymity and data havens (such as Freenet) allow unconditional free speech, as the technology guarantees that material cannot be removed and the author of any information is impossible to link to a physical identity or organization.

In November 2007, "Father of the Internet" Vint Cerf stated that he sees Government-led control of the Internet failing due to private ownership.

Some commonly used methods for censoring content are:

1) IP blocking. Access to a certain IP address is denied. If the target Web site is hosted in a shared hosting server, all websites on the same server will be blocked. This affects IP-based protocols such as HTTP, FTP and POP. A typical circumvention method is to find proxies that have access to the target websites, but proxies may be jammed or blocked, and some Web sites also block proxies. Some large websites like Google have allocated additional IP addresses to circumvent the block, but later the block was extended to cover the new IPs.

2) DNS filtering and redirection. Don't resolve domain names, or return incorrect IP addresses. This affects all IP-based protocols such as HTTP, FTP and POP. A typical circumvention method is to find a domain name server that resolves domain names correctly, but domain name servers are subject to blockage as well, especially IP blocking. Another workaround is to bypass DNS if the IP address is obtainable from other sources and is not blocked. Examples are modifying the Hosts file or typing the IP address instead of the domain name in a Web browser.

3) Uniform Resource Locator (URL) filtering. Scan the requested URL string for target keywords regardless of the domain name specified in the URL. This affects the HTTP protocol. Typical circumvention methods are to use escaped characters in the URL, or to use encrypted protocols such as VPN and TLS/SSL.

4) Packet filtering. Terminate TCP packet transmissions when a certain number of controversial keywords are detected. This affects all TCP-based protocols such as HTTP, FTP and POP, but Search engine results pages are more likely to be censored. Typical circumvention methods are to use encrypted connections—such as VPN and TLS/SSL—to escape the HTML content, or by reducing the TCP/IP stack's MTU/MSS to reduce the amount of text contained in a given packet.

5) Connection reset. If a previous TCP connection is blocked by the filter, future connection attempts from both sides will also be blocked for up to 30 minutes. Depending on the location of the block, other users or websites may also be blocked if the communication is routed to the location of the block. A circumvention method is to ignore the reset packet sent by the firewall.

6) Web feed blocking. Increasingly, incoming URLs starting with the words "rss", "feed", or "blog" are blocked.

7) Reverse surveillance. Computers accessing certain websites including Google are automatically exposed to reverse scanning from the ISP in an apparent attempt to extract further information from the "offending" system

8) One of the most popular filtering software programmes is SmartFilter, owned by Secure Computing in California, which has recently been bought by McAfee. SmartFilter has been used by Tunisia, Saudi Arabia and Sudan, as well as in the US and the UK.

Automatic censorship sometimes stops matter which it was not intended to stop. An example is that automatic censorship against sexual words in matter for children, set to block the word "cunt", has been known to block the Lincolnshire (UK) placename Scunthorpe.

Circumvention

There are a number of resources that allow users to bypass the technical aspects of Internet censorship. Each solution has differing ease of use, speed, and security from other options. Most, however, rely on gaining access to an internet connection that is not subject to filtering, often overseas or in a neighbouring state not subject to the same censorship laws. This is an inherent problem in internet censorship in that so long as there is one publicly accessible system in the world without censorship, it will still be possible to have an access censored material.

Proxy Websites

Proxy websites are often the simplest and fastest way to access banned websites in censored nations. Such websites work by being themselves un-banned but capable of displaying banned material within them. This is usually accomplished by entering a URL address which the proxy website will fetch and display. They recommend using the https protocol since it is encrypted and harder to block.

Java Anon Proxy

Java Anon Proxy is primarily a strong, free and open source anonymizer software available for all operating systems. As of 2004, it also includes a blocking resistance functionality that allows users to circumvent the blocking of the underlying anonymity service AN.ON by accessing it via other users of the software.The addresses of JAP users that provide a forwarding server can be retrieved by getting contact to AN.ON's InfoService network, either automatically or, if this network is blocked, too, by writing an e-mail to one of these InfoServices. The JAP software automatically decrypts the answer after the user completes a CAPTCHA. The developers are currently planning to integrate additional and even stronger blocking resistance functions.

Virtual Private Networks (VPNs)

Using Virtual Private Networks, a user who experiences internet censorship can create a secure connection to a more permissive country, and browse the internet as if they were situated in that country. Some services are offered for a monthly fee, others are ad-supported.

Psiphon

Psiphon software allows users in nations with censored Internet such as China to access banned websites like Wikipedia. The service requires that the software be installed on a computer with uncensored access to the Internet so that the computer can act as a proxy for users in censored environments.

Sneakernets

Sneakernet is a term used to describe the transfer of electronic information, especially computer files, by physically carrying data on storage media from one place to another. A sneakernet can move data regardless of network restrictions simply by not using the network at all.

The volunteer organization Information Without Borders is attempting to implement a sneakernet routing protocol for providing cheap Internet access to developing and post-conflict regions using donated flash drives, PDAs and mobile phones. The protocol is also useful for providing free and open Internet access to people living under repressive regimes that restrict free expression by limiting access. This protocol is still under development, but actual flash-drive sneakernets are known to exist in Cuba. Flash-drive sneakernets were used in 2008 to distribute a video of a student asking why Cubans are not permitted to access web sites like Yahoo.

OpenNet Initiative (ONI)

The OpenNet Initiative (ONI) is a joint project whose goal is to monitor and report on internet filtering and surveillance practices by nations. The project employs a number of technical means, as well as an international network of investigators, to determine the extent and nature of government-run internet filtering programs. Participating academic institutions include the Citizen Lab at the Munk Centre for International Studies, University of Toronto; Berkman Center for Internet & Society at Harvard Law School; the Oxford Internet Institute (OII) at University of Oxford; and, The SecDev Group which took over from the Advanced Network Research Group at the Cambridge Security Programme, University of Cambridge.

In 2006, Reporters without Borders (*Reporters sans frontières*, RSF, a Paris-based international non-governmental organization that advocates freedom of the press) published a list of the 13 "enemies of the Internet": The organization classifies a country as an enemy of the internet because "all of these countries mark themselves out not just for their capacity to censor news and information online but also for their almost systematic repression of Internet users." The list is updated annually and now includes 12 countries in 2009:

1. Burma
2. China
3. Cuba
4. Egypt
5. Iran
6. North Korea
7. Saudi Arabia
8. Syria
9. Tunisia
10 Turkmenistan
11 Uzbekistan
12 Vietnam

While there is no universally agreed upon definition of what constitutes "pervasive censorship", RSF (*Reporters sans frontières*) maintains an internet enemy list while the OpenNet Initiative categorizes some nations as practicing extreme levels of Internet censorship. Such nations often censor political content and may retaliate against citizens who violate the censorship with imprisonment or other sanctions.

Burma

Burma is in ONI's substantial category and is on the RSF's internet enemy list. Internet use is so threatening in Burma that even going online is considered a dissident act, thus giving the country a very low penetration rate. The internet is regulated by the Electronic Act which bans the importing and use of a modem without official permission, and the penalty for violating this is a 15 year prison sentence, as it is considered "damaging state security, national unity, culture, the national economy and law and order."

Burma has banned the websites of political opposition groups, sites relating to human rights, and organizations promoting democracy in Burma. During the 2007 anti-government protests, Burma completely shut down all internet links from its country.

Cuba

Cuba is on ONI's watchlist and on RSF's internet enemy list. According to Reporters Without Borders, Cuba has the lowest ratio of computers per inhabitant in Latin America, and the lowest internet access ratio of all the Western hemisphere. Citizens have to use government controlled "access points", where their activity is monitored through IP blocking, keyword filtering and browsing history checking. The government cites its citizens' access to internet services are limited due to high costs and the American embargo, but there are reports concerning the will of the government to control access to uncensored information both from and to the outer world. The Cuban government continues to imprison independent journalists for contributing reports through the Internet to web sites outside of Cuba.

Salim Lamrani, a professor at Paris Descartes University, has accused Reporters Without Borders with making unsupported and contradictory statements regarding Internet connectivity in Cuba. However, despite precise figures are hard to know because of the secretive nature of the regime and its telecommunication policies, testimonials from independent bloggers, activists and international watchers agree by saying the difficulties for non-government people to access the web as well as the harsh punishments against transgressors are the norm. The Committee to Protect Journalists has pointed Cuba as one of the ten most censored countries around the world.

Egypt

Egypt is not categorized by the ONI and is on the RSF's internet enemy list. Due to fears of terrorism, the government increased web surveillance in 2007. To connect to wireless internet in a public place, such as a cybercafé, a person must give up a lot of personal information, such as a phone number or ID #, making it hard for citizens to express themselves freely.

Iran

Iran is in ONI's pervasive category and on RSF's internet enemy list. Iran Internet censorship is delegated to ISPs who attempt to filter contents critical of the government, pornographic websites, political blogs, and especially recently women's rights websites, weblogs, and online magazines. Bloggers in Iran have been imprisoned for their Internet activities. Most recently, the Iranian government temporarily blocked access, between 12 May 2006 and January 2009, to video-upload sites such as YouTube.com. Flickr, which was blocked for almost the same amount of time was opened in February 2009. But after 2009 election protests YouTube, Flickr, Twitter, Facebook and many more websites got blocked again.

North Korea

North Korea is not categorized by ONI but is on RSF's internet enemy list. Only a few hundred thousand citizens in North Korea, representing about 4% of the total population, have access to the Internet, which is heavily censored by the national government. According to the RSF North Korea is a prime example where mediums of communication are controlled by the government. According to the RSF, Internet is the medium most used to the service of the North Korean government to primarily spread propaganda. The North Korean network is monitored heavily with only two websites being hosted under a domain name.

People's Republic of China

The People's Republic of China is in ONI's pervasive category and is on RSF's internet enemy list. China blocks or filters Internet content relating to Tibetan independence, Taiwan independence, police brutality, the Tiananmen Square protests of 1989, freedom of speech, pornography, some international news sources (such as the VOA), certain religious movements (such as Falun Gong), and many blogging websites. Some 52 cyber dissidents are reportedly imprisoned in China for their online postings.

Syria

Syria has banned websites for political reasons and arrested people accessing them, and is in ONI's pervasive category and is on RSF's internet enemy list. In addition to filtering a wide range of Web content, the Syrian government monitors Internet use very closely and has detained citizens "for expressing their opinions or reporting information online." Vague and broadly worded laws invite government abuse and have prompted Internet users to engage in self-censoring and self-monitoring to avoid the state's ambiguous grounds for arrest.

Tunisia

Tunisia is in ONI's pervasive category and is on RSF's internet enemy list. Tunisia has blocked thousands of websites (such as pornography, mail, search engine cached pages, online documents conversion and translation services) and peer-to-peer and FTP transfer using a transparent proxy and port blocking. Cyber dissidents including pro-democracy lawyer Mohammed Abbou have been jailed by the Tunisian government for their online activities.

Turkmenistan

Turkmenistan is not categorized by the ONI and it is on the RSF's internet enemy list. Internet usage in Turkmenistan is under tight control by the government. Turkmen got their news through satellite television until 2008 when the government decided to get rid of satellites leaving Internet as the only medium where information could be gathered. Internet is monitored thoroughly by the government as websites ran by human rights organizations and news agencies were blocked.

Uzbekistan

Uzbekistan is in ONI's pervasive category and is on RSF's internet enemy list. Uzbekistan prevents access to websites regarding banned Islamic movements, independent media, NGOs, and material critical of the government's human rights violations. Some Internet cafes in the capital have posted warnings that users will be fined for viewing pornographic websites or website containing banned political material.

Vietnam

Vietnam is in ONI's pervasive category and is on RSF's internet enemy list.

The main networks in Vietnam prevent access to websites critical of the Vietnamese government, expatriate political parties, and international human rights organizations, among others. Online police reportedly monitor Internet cafes and cyber dissidents have been imprisoned for advocating democracy.

Bahrain

On 5 January 2009 the Ministry of Culture and Information issued an order (Resolution No 1 of 2009) pursuant to the Telecommunications Law and Press and Publications Law of Bahrain that regulates the blocking and unblocking of websites. This resolution requires all ISPs—among other things—to procure and install a website blocking software solution chosen by the Ministry. The Telecommunications Regulatory Authority ("TRA") assisted the Ministry of Culture and Information in the execution of the said Resolution by coordinating the procurement of the unified website blocking software solution. This software solution is operated solely by the Ministry of Information and Culture and neither the TRA nor ISPs have any control over sites that are blocked or unblocked.

South Korea

South Korea is in ONI's substantial category but is not on RSF's internet enemy list. South Korea's internet censorship policy is highly political and particularly strong toward suppressing anonymity in the Korean internet. In 2007, numerous bloggers were censored and their posts deleted by police for expressing criticism of, or even support for, presidential candidates. This even lead to some bloggers being arrested by the police. Subsequently in 2008, just before a new presidential election, new legislation that required all major internet portal sites to require identity verification of their users was put into effect. This applies to all users who add any publicly viewable content. For example, to post a comment on a news article, a user registration and citizen identity number verification is required. For foreigners who do not have such numbers, a copy of passport must be faxed and verified. Although this law was initially met with public outcry, as of 2008, most of the major portals, including Daum, Naver, Nate, and Yahoo Korea, enforce such verification before the user can post any material that is publicly viewable

Saudi Arabia

Saudi Arabia is in ONI's substantial category and is on RSF's internet enemy list. Saudi Arabia directs all international Internet traffic through a proxy farm located in King Abdulaziz City for Science & Technology. Content filtering is implemented there using software by Secure Computing. Additionally, a number of sites are blocked according to two lists maintained by the Internet Services Unit (ISU): one containing "immoral" (mostly pornographic) sites, the other based on directions from a security committee run by the Ministry of Interior (including sites critical of the Saudi government). Citizens are encouraged to actively report "immoral" sites for blocking, using a provided Web form. The legal basis for content-filtering is the resolution by Council of Ministers dated 12 February 2001. According to a study carried out in 2004 by the OpenNet Initiative:

United Arab Emirates

The United Arab Emirates is in ONI's substantial category and is not on Reporters_Without_Borders (RSF)'s internet enemy list. The United Arab Emirates forcibly censors the Internet using Secure Computing's solution. The nation's ISPs Etisalat and du (telco) ban pornography, politically sensitive material, and anything against the perceived moral values of the UAE. All or most VoIP services are blocked.

Yemen

Yemen is in ONI's substantial category and is not on RSF's internet enemy list. Yemen's two ISPs block access to contents falling under the categories of gambling, adult contents, and sex education as well as material seeking to convert Muslims to other religions.

Australia

Australia is in ONI's nominal category as of 2008. It does not allow content that would be classified "RC" (Refused Classification or banned) or "X18+" (hardcore non-violent pornography) to be hosted within Australia and considers such content "prohibited"/"potentially prohibited" outside Australia; it also requires most other age-restricted content sites to verify a user's age before allowing access. Since January 2008 material that would be likely to be classified "R18+" or "MA15+" and which is not behind such an age verification service (and, for MA15+, which also meets other criteria such as provided for profit, or contains certain media types) also fits the category of "prohibited" or "potentially prohibited". The regulator ACMA can order local sites which do not comply taken down, and overseas sites added to a blacklist provided to makers of PC-based filtering software.

Belgium

Belgian internet providers Belgacom, Telenet, Base, Scarlet, EDPnet, Dommel, Proximus, Mobistar, Mobile Vikings, Tele2, and Versatel have started filtering several websites on DNS level since April 2009. People who browse the internet using one of these providers and hit a blocked website are redirected to a page that claims that the content of the website is illegal under Belgian law and therefore blocked.

Brazil

Brazilian legislation restricts the freedom of expression (Paim Law), directed especially to publications considered racist (such as neo-nazi sites). The Brazilian Constitution also prohibits anonymity of journalists.

Canada

Canada is in ONI's nominal category and is not on RSF's internet enemy list. In a few cases, information which the government is actively attempting to keep out of Canadian broadcast and print media (such as names of young offenders or information on criminal trials subject to publication bans) is available to Canadian users via Internet from sites hosted outside Canada.

Chile

Chile is not categorized by ONI and is not on RSF's internet enemy list and it's considered one of the most liberal countries in terms of internet freedom in Latin America. Many educational institutions (universities and schools) block the access to websites like YouTube, Fotolog, Flickr, Blogger, Rapidshare, Twitter and Facebook, depending of the institution; in some cases also popular portals like Terra.cl, LUN.com, EMOL.com are also blocked; pornography, specially any kind of child pornographic website is blocked The Chilean Government also block the access in their computers to blogs or electronic versions of the local newspapers with opinions against the Government or the ruling coalition, for example, during the first days of Transantiago or the 2006 Student Protests or the Chile-Microsoft relationship.

Colombia

Colombia blocks several websites as part of its Internet Sano program.In December 2009, an internaute was sent to prison for promoting a group against president Alvaro Uribe's sons.

Czech Republic

Since 2008, mobile operators T-mobile and Vodafone pass mobile and fixed Internet traffic through Cleanfeed which uses data provided by the Internet Watch Foundation to identify pages believed to contain indecent photographs of children, and racist materials.

Denmark

Denmark is not categorized by ONI and is not on RSF's internet enemy list. Denmark's biggest Internet service provider TDC A/S launched a DNS-based child pornography filter on 18 October 2005 in cooperation with the state police department and Save the Children, a charity organisation. Since then, all major providers have joined and as of May 2006, 98% of the Danish Internet users are restricted by the filter. The filter caused some controversy in March 2006, when a legal sex site named Bizar.dk was caught in the filter, sparking discussion about the reliability, accuracy and credibility of the filter. Also, as of 18 October 2005, TDC A/S has blocked access to AllOfMP3.com, a popular MP3 download site, through DNS filtering.

Fiji

Fiji is not categorized by ONI and is not on RSF's internet enemy list. In May 2007 it was reported that the military in Fiji had blocked access to blogs critical of the regime.

Finland

Finland is not categorized by ONI and is not on RSF's internet enemy list. Following a "voluntary law" enacted by Finnish parliament in 1 January 2007, most of the Finland's major Internet service providers decided on 22 November 2006 to begin filtering child pornography and ISPs first started filtering on January 2008. The Ministry of Communications has commented that filtering is voluntary for ISPs as long as they do not refuse. The blacklist is provided by Finnish police and should contain only foreign sites. Technically filtering was planned to be URI based like the United Kingdom's Cleanfeed, but so far implementations have been DNS based.

France

France is in ONI's watchlist and is not on RSF's internet enemy list. French courts demanded Yahoo! block Nazi material in the case LICRA vs. Yahoo. The case is currently on appeal for an en banc rehearing.

Ghana

Ghana is not categorized by ONI and is not on RSF's internet enemy list. In 2002 the government of Ghana censored internet media coverage of tribal violence in Northern Ghana.

India

India is in ONI's nominal category and is not on RSF's internet enemy list. As of July 2006 the Indian government has directed ISPs to block seventeen websites, including some hosted on the Geocities, Blogspot and Typepad domains. Initial implementation difficulties led to these domains being blocked entirely. Access to sites on these domains other than the specifically banned ones was restored by most ISPs after about a week. The first documented incident of Internet censorship in India was the Yahoo! Groups ban of 23 September 2003. Kynhun, a Yahoo! group linked to the outlawed "Hynniewtrep National Liberation Council", a minor separatist group, was ordered banned by the Department of Telecommunications. Difficulties in implementing the ban by the ISP's ultimately led to all Yahoo! groups being banned for a period of about two weeks.Recently, Indian law enforcement has entered an agreement with the popular social networking site Orkut to track down what it deems defamatory content which, in their example, includes content critical of Bal Thackeray. India is also looking to block Google Maps due to the Mumbai attacks.

Israel

Israel is not categorized by ONI and is not on RSF's internet enemy list. The Orthodox Jewish parties in Israel proposed an internet censorship legislation would only allow access to adult-content Internet sites for users who identify themselves as adults and request not to be subject to filtering. In 27/02/2008 the law passed in its first of three votes required, however it has been rejected by the government's legislation committee on 12/07/2009. Israeli laws regarding the holocaust denial apply to the Internet as well as to traditional media. Sites which deny the holocaust cannot be hosted in Israel and Israelis who attempt to use the Internet for the purposes of holocaust denial face arrest and imprisonment.

Italy

Italy is not categorized by ONI and is not on RSF's internet enemy list. Italy bans the use of foreign bookmakers over the Internet by mandating certain edits to DNS host files of Italian ISPs. Italy is also blocking access to websites containing child pornography. In 2008, Italy blocked also The Pirate Bay website for some time, basing this censorship on a law on electronic commerce.

Jordan

Jordan is not on RSF's internet enemy list and censorship is relatively light. Access to Internet content in the Hashemite Kingdom of Jordan remains largely unfettered, with filtering selectively applied to only a small number of sites. However, media laws and regulations encourage some measure of self-censorship in cyberspace, and citizens have reportedly been questioned and arrested for Web content they have authored. Censorship in Jordan is mainly focused on political issues that might be seen as a threat to national security due to the nation's close proximity to regional hotspots like Israel, Iraq, Lebanon, and the Palestinian territories. Jordan, unlike most of its neighbors, has a free and an advanced telecommunications sector.

Mexico

In May 2009, the Mexican Federal Electoral Institute (IFE), asked Youtube to remove a parody of Fidel Herrera, governor of the state of Veracruz. Negative advertising in political campaigns is prohibited by present law, although the video appears to be made by a regular citizen which would make it legal. It was the first time a Mexican institution intervened directly with the Internet. However, Mexico has yet to censor any sites.

Morocco

Morocco is in ONI's watchlist and is not on RSF's internet enemy list. As of March 2006, Morocco had blocked access to a few blogging sites, such as LiveJournal. Reporters Without Borders says that Morocco now censors all political websites advocating Western Sahara's independence, however many Western Sahara-related blogs are still accessible. Google Earth has also been added to the list of filtered Web sites in Morocco. In 2007 Morocco's main telecommunication operator Maroc Telecom also blocked access to YouTube for nearly a month, without giving any reason.

Netherlands

Since 2007 in the Netherlands one major ISP, UPC Netherlands, blocks access on DNS level to sites authorities claim are known to provide child pornography. In the second quarter of 2008 all major Dutch ISPs have agreed with Ernst Hirsch Ballin of the Ministry of Justice to also block all the sites that are on the list. The blacklist is compiled by the National Police Forces (KLPD). Ernst Hirsch Ballin has said that at the moment 150 websites are blocked. It contains no websites that are hosted in EU countries and they are checked once every 2 months by Productteam Bestrijding Kinderpornografie. Providers will not be forced to use it since that would be unconstitutional according to a research done by the governmental Scientific Research- and Documentation Center (WODC) commissioned by the Ministry of Justice. As of 2009 the only providers that use the filter are UPC and two small providers, Scarlet and Kliksafe. The providers that have been positive about a non-mandatory filter do not have it in use.

Norway

Norway is in ONI's watchlist and is not on RSF's internet enemy list. Norway's major Internet service providers have a DNS filter which blocks access to sites authorities claim are known to provide child pornography, similar to Denmark's filter.

Pakistan

Pakistan is in ONI's watchlist and is not on RSF's internet enemy list. Pakistan has blocked access to websites critical of the government. Currently, the government has blocked blogs hosted on Blogspot.com. A ban on pornographic websites has also been enacted.. Websites promoting Balochistan and Sindh nationalism or documenting human rights abuses in those places have also been banned.

Poland

Although in Poland the Internet is basically free from state interventions there is a problem with portal censorship. Most popular forums are heavily manipulated by their managements through organized mass-posting of preferred views. This kind of manipulation may hamper discussions between people of different outlooks and falsify the representativeness of a forum.As of December 2008, Poland's main ISP,TPSA, started to occasionally block some websites, which they deem "improper", most notably child pornography and adult sites, but they also block some IRC channels and they used to block gimp.org website for some time.

At present some legalisations about Internet censorships are proposed. There are propositions to build Register of blocked web sites, based on government organ choice (Police, Ministry of Finance, Secret Police etc.). In project administrative decisions, without court intervention, may be performed, and then chosen address would be blocked. Owner of the site would not be informed until performing blockade. Then may be possible to follow procedure to legalise blocked site.

Russia

Russia is in ONI's watchlist and is not on RSF's internet enemy list. Russia pressured Lithuania into shutting down the Kavkaz-Center website, a site that hosts videos on attacks on Russian forces in Chechnya, and reports on the Second Chechen War from a Chechen separatist perspective. In February 2008, it became known that six Russian internet providers with ties to the government were blocking access to an opposition aggregate news site. After this became public, the biggest of these companies dropped the block and explained that it was "testing content filters". The other five blocks remain in place.

Singapore

Singapore is in ONI's nominal category and is not on RSF's internet enemy list. In Singapore, three people were arrested and charged with sedition for posting racist comments on the Internet, of which two have been sentenced to imprisonment. Some ISPs also block internet content related to recreational drug use. Singapore's government-run Media Development Authority maintains a confidential list of blocked websites that are inaccessible within the country. The Media Development Agency exerts control over Singapore's three ISPs to ensure that blocked content is entirely inaccessible.

Sweden

Sweden is not categorized by ONI and is not on RSF's internet enemy list. Sweden's major Internet service providers have a DNS filter which blocks access to sites authorities claim are known to provide child porn, similar to Denmark's filter. A partial sample of the Swedish internet censorship list can be seen at a Finnish site criticizing internet censorship. The Swedish police are responsible for updating this list of forbidden Internet sites. On 6 July, Swedish police said that there is material with child pornography available on torrents linked to from the torrent tracker site Pirate Bay and said it would be included in the list of forbidden Internet sites. This, however, did not happen as the police claimed the illegal material had been removed from the site. Police never specified what the illegal content was on TPB. This came with criticism and accusations that the intended The Pirate Bay's censorship was political in nature.

Thailand

Thailand is in ONI's nominal category and is not on RSF's internet enemy list. Significant efforts have been made in Thailand to oppose sites that are representing illegal activities. Activities such as gambling, drug usage and pornography are strictly banned, using DNS control in Thailand and, more effectively, a transparent proxy. This makes the website appear to be inaccessible. Also, the government has banned sites that discuss circumventing Internet censorship.

Turkey

Many minor and major websites in Turkey have been subject to censorship until now, including the oldest and most popular Turkish social blogging community Sourtimes and widely popular poetry and literature community Antoloji.com. On 6 March 2007, the government of Turkey blocked access to the video-upload site YouTube.com, with the following statement parked on the domain: "Access to www.youtube.com site has been suspended in accordance with decision no: 2007/384 dated 06.03.2007 of Istanbul First Criminal Peace Court." The ban was met with widespread protests and lifted two days later. Youtube was banned again in 12.03.2008 with decision no 2008/251, which was then soon lifted. As of August 2008, Youtube is still banned in the country since 5th of May 2008, due to two court decisions.

United Kingdom

The United Kingdom is in ONI's watchlist and is not on RSF's internet enemy list. British Telecommunications ISP passes internet traffic through a service called Cleanfeed which uses data provided by the Internet Watch Foundation to identify pages believed to contain indecent photographs of children. When such a page is found, the system creates a 'URL not found page' error rather than deliver the actual page or a warning page. Other ISPs use different systems such as WebMinder. Regarding CleanFeed, the list of banned websites is provided by Internet Watch Foundation (IWF). CleanFeed is a silent content filtering system, which means that internet users cannot state if they are being regulated by CleanFeed or facing connection failures. According to a small-sample survey being conducted at 2008 by Nikolaos

Koumartzis, an MA researcher at London College of Communication, the vast majority of UK based internet users (90.21%) were unaware of the existence of CleanFeed software. Moreover, 60.87% of the participants stated that they don't trust BT and 65.22% of them that don't trust IWF either to be responsible for a silent censorship system in UK.

United States of America

In 1996 the United States enacted the Communications Decency Act, which severely restricted online speech that could potentially be seen by a minor – which, it was argued, was most of online speech. Free speech advocates, however, managed to have most of the act overturned by the courts. The Digital Millennium Copyright Act criminalizes the discussion and dissemination of technology that could be used to circumvent copyright protection mechanisms, and makes it easier to act against alleged copyright infringement on the Internet. Many school districts in the United States frequently censor material deemed inappropriate for the school setting. In 2000, the U.S. Congress passed the Children's Internet Protection Act (CIPA) which requires schools and public libraries receiving federal funding to install internet filters or blocking software. Congress is also considering legislation to require schools, some businesses and libraries to block access to social networking websites, The Deleting Online Predators Act. Opponents of Internet censorship argue that the free speech provisions of the First Amendment bars the government from any law or regulation that censors the Internet.

Implications of E-Government

In the new economy, underpinned by revolutionary changes in science and technology, information and knowledge has become a key factor in economic competitiveness. Developing countries must pursue a more active role in the formulation of national policies and strategies to promote the information economy, to reap huge benefits in terms of economic and social growth/ development. eGovernment is believed to play a fundamental role to this end. It does not only facilitate market- led initiatives but it also plays a major role in initiating the process of capability building and in coordinating the actions of a large number of interested stakeholders. It offers the potential of reshaping the public sector activities and processes, building relationships between citizens and the government, enhancing transparency, increasing government capacity and providing a "voice" for those outside the gove rnment.

Nevertheless, the ability of developing countries to reap the full benefits of eGovernment is limited and is largely hampered by the existence of many political, social and economic hindrances. But, despite the barriers and impediments they experienced, the reference cases provided here show that developing countries should and could take advantages of the ICT revolution. The case analysis reveals some important tips to consider for a successful design and implementation of eGovernment initiatives.

- E-readiness assessment - Start with an e-readiness assessment study which permits stakeholders to understand the current state of telecommunication networks infrastructure, legal and regulatory framework, current level of human resources and skills as well as the main impediments within a country's borders. Based on the outcomes of this assessment, it is possible to produce strategies and action plans for building human resource capability, legislative frameworks, institutional infrastructures, technological infrastructures and accessibility for all in a tailored and effective way.
- Raise awareness among public and private organizations - Organize workshops, events, seminars, conferences with the objective of raising awareness about real opportunities and benefits that the ICT revolution can bring. Prepare for long-term solutions to problems by ensuring the availability of appropriate training programs for future management of technological and business changes.
- Think small, be agile and fast - Begin with feasible pilot projects, tailored to specific contexts. Build up steadily the qualifications necessary for facing hindrances. Be prepared to make the required changes on the road. Agility and flexibility assure the success.
- Stimulate collaboration and coordination among government departments and agencies to increase efficiency and effectively in process handling. Address challenges and opportunities in strong partnership with private organizations, major donors, research institutions and universities, and support cross fertilization of ideas, solutions and knowledge.
- Invest in human development – the success of e-initiatives depends largely on human skills and capabilities. Accordingly, education and training initiatives must be considered as priority actions. Staff need to be trained to handle new processes and activities; they have to be given incentives (not necessarily monetary) to prevent the brain drain

of skilled people; and they need to feel part of the organization by engaging in the decision making process. Some basic training needs necessary to be provided to community members, in general, in order for them to be able to use new facilities for accessing electronic information and services.

- Show sensitivity to local realities by assessing and evaluating different alternatives, ways and solutions for digital government development including mobile telephones, kiosks, and multi- channel access to services. Find viable solutions to ensure the effective participation of the community in the information economy.
- Adopt a holistic and comprehensive approach, with clear vision and strategy to overcome the barriers and challenges for change. Integrate eGovernment with other development strategies and policies to ensure a broad base diffusion. The active role of top leaders is crucial especially at the earliest stages, to raise awareness, make ICT development a national priority, build and maintain wide commitment and involvement at public and private levels.
- Prepare to manage knowledge and change - Establish knowledge management processes and tools to ensure storage, usage, easy retrieval of strategic information and knowledge for better and fast decision-making process, for further adaptation and development, for realizing the necessary improvements and always search for better and innovative value added services and solutions.

References

Dutch ICT and Government Advisory Committee (2001) *Citizen and Government in the Information Society: The Need for Institutional Innovation.* The Hague: Dutch Ministry of the Interior and Kingdom Relations.

Hirst, P. and Norton, M. (1998) *Electronic Government. Information Technologies and the Citizen,* United Kingdom Parliament Parliamentary Office of Science and Technology, http://www.parliament.uk/post/egov.htm

Osborne, D. and Gaebler, T. (1992) *Reinventing Government: How the Entrepreneurial Spirit is Transforming the Public Sector*, Addison-Wesley: MA.

Riley, B.T. (2001) *Electronic Governance and Electronic Democracy: Living and Working In The Connected World, Vol. 2*, Commonwealth Centre For Electronic Governance, Brisbane, Australia.

Bibliography

Androutsellis-Theotokis, et.al. (2010). "Open source software: A survey from 10,000 feet" (PDF). *Foundations and Trends in Technology, Information and Operations Management* 4 (3-4): 187–347.

Arendt, H. (1958) *The Human Condition.* Chicago: University of Chicago Press.

Ashok Agarwal and V Ventaka Ramana. *Foundations of e-Government.* ICEG'07 5th International Conference on e-Governance

Becker, T. and C. Slaton (2000), *The Future of Teldemocracy,* London.

Bellamy and Taylor, "Informatization and new public management," *Public Policy and Administration,* p. 29.

Bellamy and Taylor, "Informatization and new public management," *Public Policy and Administration,* p. 29.

Berry, D M.& Moss, Giles (2006). *Free and Open-Source Software: Opening and Democratising e-Government's Black Box.* Information Polity Volume 11. (1). pp. 21–34

Blumler, J.G. and S. Coleman (2001), *Realising Democracy Online: A Civic Commons in Cyberspace,* London.

Borins and Wolf, "Realizing the Potential of Public Sector Information Technology," in [Thomas, d'Ombrain, Borins and Wolf], *Change, Governance and Public Management,* p. 182.

Calhoun, C. (ed.) (1992a) *Habermas and the Public Sphere.* Cambridge, MA: MIT Press.

Castells, M. (1996), *The Rise of the Network Society, The Information Age: Economy, Society and Culture*, Vol. I, Oxford.

Chris DiBona and Sam Ockman and Mark Stone, ed. (1999). *Open Sources: Voices from the Open Source Revolution.* O'Reilly.

Cordella, A (2007), E-government: towards the e-bureaucratic form?, *Journal of Information Technology*, 22, 265–274.

Dutch ICT and Government Advisory Committee (2001) *Citizen and Government in the Information Society: The Need for Institutional Innovation.* The Hague: Dutch Ministry of the Interior and Kingdom Relations.

Fadi P. Deek and James A. M. McHugh (2008). *Open Source: Technology and Policy*. Cambridge University Press.

Garson, D.G. (2006). Public Information Technology and E-Governance. Sudbury, MA: Jones and Bartlett Publishers.

Hirst, P. and Norton, M. (1998) *Electronic Government. Information Technologies and the Citizen,* United Kingdom Parliament Parliamentary Office of Science and Technology, http://www.parliament.uk/post/egov.htm

Ignace Snellen, (2002). "Electronic Governance: Implications for Citizens, Politicians and Public Servants," *International Review* of *Administrative Sciences* 68, no. 2, June, pp. 189-90.

Jane E. Fountain, (2001), *Building the Virtual State: Information Technology and Institutional Change,* Washington, D.C.: Brookings Institution Press, p. 5.

Joshua Gay, ed. (2002). *Free Software, Free Society: Selected Essays of Richard M. Stallman*. GNU Press, Free Software Foundation.

Kushchu I. and Kuscu M. H. (2003). "From e-Government to m-Government: Facing the Inevitable". *The 3rd European Conference on e-Government*. pp. 253–260.

Mitchell, W. (1998) *City of Bits*. Cambridge, MA: MIT Press.

Organisation for Economic Co-operation and Development, *The e-Government Imperative,* p. 156.

Osborne, D. and Gaebler, T. (1992) *Reinventing Government: How the Entrepreneurial Spirit is Transforming the Public Sector*, Addison-Wesley: MA.

Paul H.A. Frissen, (1992). "Informatization in public administration: An introduction," *International Review of Administrative Sciences* 58, no. 3, p. 308.

Riley, B.T. (2001) *Electronic Governance and Electronic Democracy: Living and Working In The Connected World, Vol. 2*, Commonwealth Centre For Electronic Governance, Brisbane, Australia.

Rose, R. (2002), *The Internet's Impact on East Asian Governance: Openness, Impersonal Rules and Accountability*, Glasgow.

Rossel, Pierre, and Matthias Finger. "Conceptualizing e-Governance." Management (2007): 399-407.

Shoshana Zuboff, (1984), *In the Age of the Smart Machine: The Future of Work and Power* New York: Basic Books, p. 9.

Wellman, B. (ed.) (1999) *Networks in the Global Village*. Boulder, CO: Westview Press.

West, Darrell. (2008). *State and Federal Electronic Government in the United States*. The Brookings Institution. Retrieved on 2008-09-16.

White, H. (1992) *Identity and Control.* Princeton: Princeton University Press.

Yang, G. (2003) 'The Internet and the Rise of a Transnational Chinese Cultural Sphere', *Media, Culture & Society*, 25(4): 469—90.

Zolo, D. (1992), *Democracy and Social Complexity: a realist approach,* Oxford

ROLE OF CASTE IN INDIAN ADMINISTRATION

S.N. Singh

ANMOL PUBLICATIONS PVT. LTD.
NEW DELHI - 110 002 (INDIA)

ANMOL PUBLICATIONS PVT. LTD.
H.O.: 4374/4B, Ansari Road, Daryaganj
New Delhi - 110 002
Ph.: 23261597, 23278000
B.O.: No. 1015, Ist Main Road, BSK III Stage
III Phase, III Block
Bangalore - 560 085 (India)
Visit us at: www.anmolpublications.com

Role of Caste in Indian Administration

First Published, 2006

ISBN 81-261-2583-7

PRINTED IN INDIA

Published by J.L. Kumar for Anmol Publications Pvt. Ltd., New Delhi - 110 002 and Printed at Mehra Offset Press, Delhi.

Contents

Preface

For sheer survival and continuance of public administration, governance has to be the key concern. It may be worth noting that during its long journey of more than hundred years it has passed through several stages. Various theories and approaches were developed in this process of evolution as an academic discipline and a field of activity. The question of governance and the decision-making, both as an activity and process became a core concern in administrative thinking. The main exponent of decision making theory, Simon, attempted to define the parameters of this and developed his own views in which most of his debate and discussion centres around 'rationality paradigm'. Also, the 'fact' and 'value', 'means' and 'ends' and other paradigms were developed to explain the concept. In the initial stages, we find the focus on 'value neutrality' in order to achieve efficiency and economy in the administration. Subsequently, allegedly new factors started affecting the governance and the decision making process, virtually questioning the very relevance of the earlier premises. How far these new concerns have influenced to help improve the governance and the decision making have been a matter of scientific analysis for sometime.

The present book precisely addresses itself with the question of caste and its role in the governing as well as decision making process in Uttar Pradesh in order to assess the nature and the extent of changes effected. The observation and the analysis is based on the primary data collected through questionnaire administered on the civil servants of the States belonging to IAS and PCS along with some sub-ordinate service officers. The book also has made use of secondary sources of data in its analysis and formation of opinion. We, on our part, are trying to examine the decision making as an activity where the civil servants have a larger role to play along with the political leadership involved in the process at different levels of governance. We have tried to analyze the role and relevance of value-neutrality premise along with value-loaded caste factor influence the decision making in the State and try to assess its impact and relevance in contemporary social-political ethos. We feel the administration has to address itself to the larger societal interest and goals. In this process the role that caste plays influences and attacks at the very base of value-neutral governance. In the larger interest we thought to probe. The present work is the cultivation of that. It is hoped that the book will generate some debate and serious thinking in this area of decision making and governance for larger public good.

A number of people have helped in the pursuit of this work. It is my pleasant duty to gratefully acknowledge the help, cooperation and support extended. I express my thanks to the Indian Council of Social Science Research, New Delhi for the financial assistance provided for this study which became the base for the book, without which it would have been difficult to undertake this sensitive task. I am grateful to the then Vice Chancellor of Lucknow University, Prof. D.P. Singh for allowing me to take up this research. I express my sincere thanks to Prof. C.P. Barthwal, Prof. & Head, Department of Public Administration, Lucknow University for providing me all the help, advice and guidance in the course of this work. I express my thanks to my colleagues in the Department, Prof. Kumkum Kishore, Dr. Manoj Dixit, Dr. Vaishali Saxena and Dr. N.L. Bharti for their valued opinions and persistent interest in things academic which has been a source of inspiration for all these years. I am equally thankful to Dr. K.K. Srivastava, Dr. S.S. Chauhan, Dr. Anand Sachan who helped me a lot in the process of data collection. I express my thankfulness to the members IAS, PCS and Subordinate Service who were kind enough to spare some time for this work and their valued opinions. A word of special thanks for the departmental Library staff, particularly, Mr. Y.P. Singh and Mr. U.S. Srivastava for their ever-willing support and assistance. I am thankful to Mr. Pramod Kumar for getting the Ms computer typed in record time. I am grateful to my wife and children for their continuous encouragement and the quiet support for all my ventures even at the cost of pains of isolation. At the end I express my deep sense of gratitude to all those who helped me in this effort whose names do not figure.

S.N. Singh

Chapter 1
Prologue

The interlinking of caste and governance has been part of India's political history. The state in ancient India functioned in a society which was hierarchical in its ordering and authority. This hierarchy was determined by the rules of the caste system. A glance at the recorded history shows that India persisted with a hierarchical, heterogeneous social order for centuries. Various authoritative texts, from ancient to modern, provide testimony to the prevalence of the caste as a social system. By implication it was a part of governing process, in the sense that many political and administrative decisions were influenced by this caste factor. The impact of the caste factor on the governing process, however, could be a matter of opinion and debate.

The administration, more so the public administration during its more than hundred years of academic journey has passed through several stages. One of the major concerns during this period has been the decision making with which all administration is concerned. Value-neutrality premise has been the major thrust in search for a science of administration. Subsequently this premise was questioned and search for a new premise was on. During the continued search for efficiency in public administration, several variables have entered the scene. In view of the importance of decision-making as an activity and as a process in governance, it is pertinent to probe into this area of state activity. Of late, caste as a factor has taken primacy over other factors in administrative process.

We propose to analyze caste, as a concept and its role in the governing process before taking up other related issues.

There is considerable focus on caste to explain the societal process world over. The term caste has been widely used to describe ranked groups within rigid system of social stratification, and specially those which constitute the society of Hindu India. A.L. Kroebar defined caste as an endogamous and hereditary sub-division of an ethnic which occupying a position of superior or inferior rank of social esteem in comparison with other such sub-division (1930- p.254). Thirty five years later and with many times as much research literature available on India and social stratification, the definition has not been significantly improved upon, although there has been greatly increased understanding of both of the Indian caste system and of other systems of stratification. A more comprehensive approaches, which make cross-cultural comparability without sacrificing cultural content can be derived from three bodies of descriptive and analytical literature studies of social stratification, studies of cultural pluralism and studies of social interaction caste systems are indeed rigid systems of social stratification, but they are also best understood in terms of distinctive patterns of social interaction[1].

A caste system, then, can be said to occur when a society is composed of birth ascribed hierarchical order and culturally different groups (caste). The hierarchy entails differential evaluation, differential rewards and differential association. Caste are ranked ultimately in terms of the shared worth, and that is ascribed by worth to the individuals who constitute them. Caste system rank people by birth group rather than by individual attributes. Rank is occupied by differential power and other rewards contingent upon caste membership; access to goods and services and other valued things. The ability to influence the behaviour of others, the source of once livelihood, the kind and amount of food, shelter and medical care of education, justice, esteem and pleasure, all these things which an individual will receive during his life, and the length of life itself are determined in large measure by caste status[2].

A caste hierarchy is to a large extent on interactional hierarchy. Social interaction is, interestingly symbolic, i.e. it has meaning-Rank is expressed and validated in interaction between persons. It is manifested in patterns of inter-personal behaviour in patterns of association. It is manifested in patterns of inter-personal behaviour and in patterns of association, who may be one's friend, wife, neighbor, master, servant, client or competitor is largely a matter of caste. Only within the caste, status equality is found[3]. Between caste, any kind of interaction which defines or jeopardize the value of hierarchy is taboo, even when such behaviour does not directly challenge the official basis of the rank system. It is becoming increasingly evident that caste cannot be defined adequately without reference to patterns of interaction. An interactional definition of a caste system might be a system of birth ascribe groups each of which constitutes for its members in the maximum limit of status equal interaction and between all of which interaction is consistently hierarchical. A caste then might be defined as a network of status equal interaction in a society characterized by network of hierarchical network between birth ascribed groups. The interaction range from informal social encounter to marriages and include wide variety of networks, such as these based on occupation, economics, politics, rituals and friendship[4]. Understanding hierarchical interaction between caste in the existence of what has been tarred "status summation" (Barth, 1960, pp. 144, 66). The multiple roles played by individual members of a caste are equivalent in the status they confer. Thus a person of high ritual status tends also to be of high economic, political and social status. These statuses tend to coalesce and people are enabled as well as enjoined to interact with members of other caste in an unambiguous, consistent and hierarchical manner[5].

CASTE AND OTHER SOCIAL ORGANIZATIONS

Caste system resembles plural societies in the cultural

distinctiveness and in the dimension found among caste on many key values and attitudes, for cultural plurality obtains when two or more different cultural traditions characterizes the population of a given society (Smith, 1965, p. 14). This pinpoints an important reference between caste and non-caste system of stratification. Stratification theory proposes, explicitly or implicitly that among people who constitute rank continuum there is a wide consensus on the criteria and expression of rank in a society. It also implies a discontinues status order lacking any foundation in any system of common interest and values while its component sections are given status continua, distinguished by their differing system of value, action and social relations (Smith, p. 83). The plural society is held together by power rather than by consensus.

Several scholars have considered Indian society as a "Caste Society". Clearly India falls short of being plural society by criteria referred earlier of the ideal type and probably all caste systems do. Caste system especially that of India have many plural features. Several contemporary social anthropologist have made the point, for example, that while castes in India share much by way cultural traditions, values, attitudes and goals, they are culturally different, and its extent is considerably less than in societies which are culturally and socially continuous. Beteille[6] has outlined the basic features of the "caste" model of Indian society while examining its usefulness as a scheme of analysis. These features are: (i) it is based on ideas and not on the actual behaviour of the people, (ii) these ideas are found in the classified text theory (iii) the system is based in 'rules of the games', and (iv) different castes fulfill complementary function, and their mutual relations are "non-autogonistic".

It can be suggested that all caste systems are characterized by plural features similar to those found in the Indian instance and that all caste system are held together in large measuring by consideration of relative power among caste-power expressed physically, economically, politically and socially[7]. India and the United States are typical in that the caste system of each functions as a result of sanctions in the hands of dominant group(s) and is really upset if the balance of power as perceived by those in the system, changes (Berrman, 1960). The dominant caste(s) exercise the power which maintains the *status quo* as does the dominant group in a plural society[8]. Caste system combines the principles of stratification and pluralism. A caste system resembles a plural society whose discrete sections are on ranked vertically. A plural society resembles a caste system wherein the groups (except the dominant one) are unranked relative to one another. In both instances there is a dominant group who sanctions assure persistent of the system and particularly its component parts. In the village caste system of India, this is the dominant caste of castes, in plural societies, it is often the colonial power[9]. This is an argument to support that India is not a plural society.

THE INDIAN CASTE SYSTEM

Studies by Romila Thaper[10], K.M. Panikkar[11], Burtan Stenn[12] and Imtiaz Ahmad have proved that Indian society was never static. Special and status mobilities existed, and these occurred due to certain stresses and strain between the ruling caste and other castes. Thus castes change their status in due course of time after their migration to new areas. However, their migration resulted into elevation of the caste as status group. The cause of migration was not ritualistic in nature. It was more due to 'relational' problem between higher and lower caste, hence, a class analysis of the Indian caste system. Thus, at all times, ancient to till today, 'class relation' have been an integral part of the India social structure including the caste system[14]. Scholars have the stride the focus the process of the caste system through different periods of Indian history.

The Indian caste system presents an ex-ordinarily complex social phenomenon. The great size and spatial extension of the population concerned and the close-inter locking of religious and secular features have produced a luxuriance of local variation from which it is difficult to draw consistent features and give them their precise emphasis[15]. It is only in the recent years that detailed study of local caste hierarchy has been made. Hence the relation of the Indian caste system to other systems of stratification and social grouping has by no means been agreed upon.

The caste system in India has been in existence all through the recorded history in one form or the other. Even under the centralized states like that of the Mauryan states hierarchies are traced. The power hierarchies too had the basis which can be believed to be the caste organised from hiegher to lower. The social hierarchy based on the prevailing caste structure perished all along. The social reform movements in various phases of our history are reflective of the fact that some contemporary reformers tried to address themselves towards the ill effects of the rigid hierarchical caste system and the rituals associated with them. The caste controlled, however, controlled almost the whole life of man; told him what and with whom he could eat and drink. What occupation are legitimate for him and whom he would marry. Overall these written and unwritten laws, usages and conventions of social and personal behaviour, the king had little control. There were caste panchayats too who mastered the societal process.

At the time of independence, most reflective Indians felt that they had come to a watershed on the other side of which were many new possibilities, while it was up to them to seize and turn to country's advantage. They had a new Constitution, a new Five-Year Plan strategy and they felt that, given the will power, they could do many things to bring justice, equality, growth and well being for the people of India. Certain things needed to be put out of way in order to achieve these objectives, to eradicate poverty, hunger, malnutrition, illiteracy, ignorance and superstition. The caste system too was counted among the obstacles in the way of process. Certainly it could be fair to say that those who fashioned the Constitution of modern India and plan strategy did not wish to give a new lease of life to caste[16].

The Indian perception of caste at the time of independence

had been shaped, at least to some extent by the dialectic between the colonial and the nationalists' views regarding its place in the present and future society of India. For the Indian intellectuals, it had become a matter of pride to compute the colonial view that, however, eloquently they might talk about democracy and development, Indians were and would remain under the grip of caste with its exclusiveness, its hierarchy and its ineluctable argumentation of civil society. But the miss-perception of caste among Indian intellectuals was not based solely on national pride, it had another more tangible, if not credible basis. Indians could use the evidence, the colonial administrators themselves provided to argue with more plausibility, that caste was definitely irreversibly in decline[17]. How far this perception lasted is again a matter of continued debate.

Max Weber, perhaps the most outstanding comparative socialists of all time, clearly, defined the social identity of the Hindus in terms of caste, and of caste in terms of rituals. "Caste, that is, the ritual rights and duties it gives and imposes and the position of Brahmans is the fundamental institution of Hinduism. Before everything else, without caste there is no Hindu" (Weber: 1958, 29). Even those who would broadly accept this view might point to significant exception: Hinduism always had a plan for renounce and the Sanyasi might still be considered a Hindu, although without caste or clan"[18]. It may, however, be noted that the Sanyasi is not an active participant in the general societal and the governing process, and therefore, he does not represent the average Hindu person.

Caste undoubtedly exists in contemporary India, it does not, however, exist as a complete system anymore, but if the phrase be permitted as a truncated system[19]. The first and most decisive causality had been the legal basis of caste (Cirramayya, 1984). No two texts can be more divergent in their orientation to caste than the Manusmriti which we may take as a charter of Hindu Law and the Constitution of India, which may be regarded as the charter of contemporary Indian law. As can be seen, comparative sociologists, from

Weber to Dumont have dealt with the interaction between caste and Hinduism, regarding it as the social expression of Hindu religion[20]. But many of the modern proponents of Hinduism, such as Vivekanand, have attacked caste instead of defending it. There are very few contemporary proponents of Hinduism who would be prepared to argue about the decline in the legal, religious and moral basis of caste if it turns out that on such an important issue as marriage rules and practices among educated Indians in the higher occupational strata, but we are hindered in our assessment of them by the paucity of systematic data[21]. It is nevertheless observed that a position to inter-caste marriage which was earlier very vocal and violent, is less vocal and at times tolerated. In the recent interview with the Editor of the Times of India, M.N. Sriniwas, noted the changing role of caste in marriage. Many other eminent sociologists have felt that the role of caste in marriages is declining as the inter-caste marriage, inter-regional and even inter-religious marriages are on the rise (Padgaonkar, 1993).

The declining notes of caste in the social or the rituals front was more than compensated on the political front. It has played a very dominant role in the post-independent India. The universal suffrage granted to people could not provide the equality of opportunity in real sense. The right to vote may be indivisible but the social set up itself is deeply divided. In these electoral processes, therefore, caste becomes the potent factor. It is intersecting to note the comments of the Mandal Commission on this – 'what caste has lost on the ritual front, it has more than gained on the political front. The caste system has provided the political leadership with channels of communication mobilization. In view of this, it will be unrealistic to assume that the institutions of caste will perish[22]. The caste factor which appeared to be on decline in the initial years vehemently stirred the socio-political process in the country. Those who were arguing for the decline of the caste system would reconsider their view and would like to find new paradigm to explain the rise of the caste system in India.

The caste has, therefore, been at the center of our socio-political process. In view of this a brief analysis of the same is being presented before attempting the analysis of other related aspects in the context of Uttar Pradesh where we made a short study.

Scholars have tried to analyse the shifting paradigm in the political process particularly Uttar Pradesh. Philip Mason argues about the inherent rivalry between different castes who have similar base. For example a caste association of Ahir or a Jat Panchayat is primarily a vehicle for interest protection of the representative caste categories and if, at a given moment there is a mutuality of interest (either political or economic or both) it is generally co-incidental[23].

Another indicator, in fact, a most significant indicator of the political shift in Uttar Pradesh is the changing caste (class) character of the State Assembly. The Legislative Assembly constituted after the 1962 elections had 46 percent upper caste representation. In the election that followed, this domination became less pronounced[24]. In the higher echelons of bureaucracy and industry the upper caste have continued to maintain their distinct edge, but in terms of political representation, they have, through successive elections shed their once unchallenged status, 1967 was one watershed, 1989 and 1993 respectively were two other manifestations of a paradigm shift. But the more important shift took place following the abolition of the zamindari system soon after independence. That one act changed the whole basis of land relations, or a class present proprietor became the dominant factor in Uttar Pradesh[25]. Hence, there seems a definite economic basis to the political change that has been watershed.

At the macro level, there has been a pronounced shift in the character of Indian politics, and what is happening in Uttar Pradesh is only a reflection of other overall patterns. In the scenario that has developed over the years, gross level issues and caste considerations have started mattering more than by ever did in setting the political agenda. Mutually hostile castes have thus had to redefine their political parameter[26]. This process has continued for some time more in Indian political process which has seen the gradual emergence of the intermediate caste on the political arena.

Caste did not enter into politics at once with independence but it made its presence strongly felt in the first general election and increasingly with each successive election[27]. In the 1950, and 1960, the leaders of all political parties condemned in public the use of caste in politics. Politicians freely acknowledged that caste was being extensively used in politics of post-independence India, but they said it was being used by other political parties and not by their own. When it was shown that their party too was using caste, their reply was that they did not start the process but were being forced to accommodate caste in order to survive in politics or at least in Indian politics, one could not afford to be for idealistic if one wished to remain in business[28]. In order to be realistic, caste was allowed to play actual part in the political process in post-independent India. Different parties in the country take the advantage of the caste factor and at times have been critical as well.

Taking a historical perspective one finds that caste has generally played a role in the socio-political process in the country in different phase of its existence. After independence also the caste continued to play its part though anthropologists and sociologists have at times doubted its role. For several years now a trend witnessed have been observed to show the role of caste in administrative process. More especially after the Mandal Commission Report in the 1980's, new focus has emerged on the socio-political area which also affected the administrative process. This has become a cause of concern for people who value efficiency and effectiveness in organizational performance.

In analysing the state system or the governing system, one can make use of several variables to explain. We, in the present context, are confining to one variable of caste and try to see the inter-linkages between the two systems to find out as to how far these are mutually accommodative and helpful or are counter productive so far as the broader national gains are concerned. We have, in the earlier pages, attempted to explain caste as a potent factor in our social system through the ages. In the following chapter we propose to take up the issues of the governance. The inter-linking of the two systems and the impact analysis is taken up at a later stage in the book.

REFERENCES

1. *Encyclopedia of Social Sciences,* Vol. – 1, 333
2. *Ibid.,* p. 334
3. However, in India it has been observed that even within the caste equality of status is not practiced as there are sub-caste within the caste hierarchy.
4. *Ibid.* pp. 334-335
5. *Ibid.* p. 335
6. Beteille Andre, "Ideas and Interest: Some Conceptual Problems in the Study of Social Stratification in Rural India", *International Social Science Journal,* Vol. 21, No. 7, pp. 17-31
7. *Encyclopedia of Social Sciences,* op. cit., p. 335
8. *Ibid.*
9. *Ibid.*
10. Romila Thaper, *Social Mobility in Ancient Society* in R.S. Sharma (Ed.) *Indian Society: Historical Probings,* New Delhi, Peoples Publishing House, 1974, pp. 95-123.
11. K.M. Panikkar, *Hindu Society at Cross Roads,* Asia Publishing House, Bombay, 1955.
12. Burton Stein, "Social Mobility and Medieval South Indian Hindu Sects", James Silverberg; *Social Mobility in the Caste System in India,* The Hague, Moutton, 1968, pp. 78-94.
13. Imtiaz Ahmad, "Caste Mobility Movements in India", *The Indian Economic and Social History Review,* Vol. III, No. 2, June 1971, pp. 164-191.
14. K.L. Sharma, *Essays on Social Stratification,* Rawat Publication, 1980, Jaipur, p. 9.

15. The Indian Caste System: *Encyclopedia of Social Sciences,* Vol., p. 339.
16. Andre Beteille, *Caste in Contemporary India,* in C.J. Fuller (ed.) *Caste Today,* Oxford University Press, 1997, pp. 153-154.
17. *Ibid.,* p. 155.
18. *Ibid.* p. 156.
19. *Ibid.,* p. 161.
20. *Ibid.*
21. *Ibid.,* p. 163.
22. Quoted from the *Report of the Mandal Commission,* 1980.
23. Pradyot Lal, Tara Nair, *Caste Vs. Caste,* Ajanta Book International, Delhi, 1998, pp. 5-6.
24. *Ibid.* p. 6.
25. *Ibid.* pp. 6-7
26. *Ibid.* p. 8
27. Andre Beteille, *op.cit.,* p.167.
28. *Ibid.*

Chapter 2

Profile: Universe and the Respondents

The state of Uttar Pradesh and its Secretariat form the base material for this book. We have largely based ourselves upon the published material in the form of books and report etc, the secondary source. We have also collected primary data through a structured questionnaire administered on the officials[1] to get their views on the theme. This part of the book briefly describes certain features of the state as a background as well as a brief profile of the respondents in order to get an idea of the universe of the study.

Uttar Pradesh is one of the states of the Indian Union which has a long but checkered history. It is inhabititated by an ancient civilization spread over a period of more than five thousand years of recorded history. It is also a place to which Lord Rama, Lord Krishna, the mythological figures belonged, who still continue to inspire the life style and the philosophy of millions of Hindus inhabiting the state and beyond the territorial boundaries of the state. It is the place where not only Hindu but other religions too have prospered. It has seen almost five hundred years of Muslim rule followed by about two hundred years of the British. The present boundaries of the state included the twelve Sarkars of the Awadh reign of Shia Dynasty i.e. the period between 1720 to 1856, a period of great social and culture harmony as well as people- friendly governance[2]. This is also a place to which

Lord Buddha belonged and preached his divine thoughts which subsequently spread over to south and south east Asia. Also, this is a place where renouned saints like Ravidas, Kabir, Tulsi and others spread the light of the knowledge and spiritual virtue. Quite a few Sufi saints too made this place their land including Hazarat Nizamuddin Aulia, Syed Salar Ghazi to name only these. Traditionally, therefore, the State had a great historical past with a multi-religious and multi-racial fervour. However, Uttar Pradesh today does not have, much to be proud of. It has been suffering from a number of ills on account of many and varied reasons. One such factor is being analysed in the present work.

THE POPULATION

Uttar Pradesh is not only the border State, but also the largest state in terms of population. Areawise it is the fourth state in the country after M.P., Rajashthan and Maharashtra. The total population of the state as per 2001 census is reported at 16,60,52,859 out of which 8,74,66,301 as male and 7,85,86,558 as female. The rural component of the states' population is 13.15 crores as against 3.45 crores urban. The percentage of rural population of the state is 70.00 per cent where as the urban population is 30.50 per cent. Thus over percent of the state's people live in rural areas.

Comparison of decadal growth of population in Uttar Pradesh (25.80 percent) and India (21.34 per cent), shows that it is higher by percent for Uttar Pradesh. According to 2001 census, the total population of India was 1,02,70,15,247, comprising of 53,12,77,708 male and 49,57,38,168 females. The UP's share in India's total population was 16,60,52,59 i.e. 16.17 per cent. In other words, every 6.14 Indian belongs to Uttar Pradesh.

LITERACY

According to 2001 census, children upto six years of age have been treated as illiterates irrespective of the fact whether they are going to school or not. Following this principle the percentage of literacy in the state is reported at 61.03. The male literacy percentage is 76.46 where the female literacy is reported at 44.34. The figure of literacy for the country stands at 65.38 percent. The male literacy is reported at 75.85 percent whereas the female literacy is reported at 54.16 percent. The figure for literacy about the country are slightly higher than the state, but at the same time the percentage clearly shows the improvement from earlier years which could be taken as a brighter side of the picture.

COMPOSITION OF THE POPULATION

An attempt has been made to collect information on the religion and caste composition of the population to have an idea of the representation of various religious and caste groups. Although the Census of India figures do not constitute the figures caste wise, they did have an information about the religious component of the population in the State in 1981 census which is as under:

MAJOR RELIGION-WISE POPULATION IN UP

Population in Lakh

Main Religious	*Community*	*Total*	*Urban*	*Rural*	*% of Total Population*
1	2	3	4	5	6
Hindu	1981	923.70	138.60	785.10	83.30
	1991	1137.12	184.65	952.46	81.74
Muslims	1981	176.60	56.70	119.90	15.90
	1991	241.09	85.94	155.15	17.33
Christian	1981	1.60	0.90	0.70	0.20
	1991	1.99	1.23	0.76	0.14
Sikh	1981	4.60	1.50	3.10	0.40
	1991	6.75	2.21	4.53	0.48
Buddhist	1981	0.50	0.20	0.30	0.10
	1991	2.21	0.56	1.64	0.16

(Contd...)

1	2	3	4	5	6
Jains	1981	1.40	1.00	0.40	0.10
	1991	1.76	1.38	0.37	0.13
Others	1981	0.20	0.10	0.10	Nil
	1991	0.16	0.04	0.12	Nil
Total	**1981**	**1108.60**	**199.00**	**909.60**	**100.00**
	1991	**1391.12**	**276.05**	**1115.60**	**100.00**

Source: ***Census of India. 1981 and 1991***
N Stands for Negligible

After giving a brief profile of the population of the State, it is now proposed to give the background of the Secretariat of the State, the main focus of the study. It is the seat of the government and the decision-making body too generally resides in the Secretariat. Hence this brief profile.

THE SECRETARIAT

The sample of the study was generally drawn from the Secretariat of the State of Uttar Pradesh. In view of this, it is proposed to present here a brief historical perspective of the same.

The U.P. Secretariat was first established formally during the British period, when a Lieutenant Governor was appointed in 1801 at Bareilly, he was allowed a Secretary and some clerks. Subsequently another Secretary was appointed in 1834 or there about but not without some initial opposition from the Supreme Government at Calcutta. Much of the local records were destroyed in fire which broke out in 1882 in Allahabad and there is very little material available on the early organization of the Secretariat. The earliest proceedings which are available and which were originally maintained at Fort William at Calcutta are for the year 1836 and 1837. They relate to Foreign and Home Department of the East India Company. By 1860, the following departments had come into existence in the Secretariat: Revenue, General, Political Judicial

(Criminal), Judicial (Civil), Police Financial and separate Revenue (Assessed Taxes license tax, excise and stamps)[3]. In course of time following additional departments were created: Scarcity, Educational, Sanitation Medical, Revenue (forest), Municipal, Infanticide, Appointment, Local Self-Government, Legislature and Miscellaneous. The Lieutenant Governor used to spend a portion of each year in the hills at Nainital and also used to move about the Province with his staff in the cold weather. The Secretariat office was accordingly divided into two branches, Head Office at Allahabad and the Camp Office which followed the Lieutenant Governor's progress through the province[4].

The first recoganisation of the State Secretariat was undertaken in 1883 and some of the important changes made in the following years that two additional Secretaries, designated as Financial Secretary and Judicial Secretary were added to relieve the Chief Secretary of the excessive work. The distribution of departments under each was as follows:

i.	Chief Secretary	Revenue (including Scarcity), Political general (including appointment) Education.
ii.	Financial Secretary	Financial, Local Self-Government, Municipal Medical & Sanitation, separate revenue.
iii.	Judicial Secretary	Judicial (Civil), Judicial (Criminal), Police, Infanticide, Oudh Revenue, Revenue (Forest)

In addition, Chief Engineer, Irrigation and Chief Engineer, Buildings and Roads Branches of the Public Department also functioned as Secretaries[5]. Office work was divided in routine, e.g. maintenance of register and copying and work of higher importance, e.g. making of precis and preparing drafts. Office establishment was accordingly divided into lower and upper section. Before 1884 all correspondence was kept weekly and monthly bundles folded and tied up between wooden pallets. This arrangement worked satisfactorily as long as correspondence was limited and references were not required to be made frequently but with increase in correspondence and multiplicity of subject and cases, it was decided to introduce subject file. The departments were made self-contained under the charge of superintendents for all stages of work connected with records, drafts and issues etc. Much of the present day secretarial system is based on that recoganisation. A system of printing of government proceedings was evolved so that duplicates of files could be available for reference at the Camp Office at Nainital during the Lieutenant Governor's stay there. Registrars were appointed in 1897. One for each Secretary's branch for general supervision and over all control on the work and staff of the departments under each Secretary. A manual of office procedure of the Civil Secretariat United Provinces of Agra and Oudh was also published in 1903[6].

When the 'Reforms' under the Government of India Act 1919, were introduced in January 1921, the transfer of numerous subjects, such as Agriculture, Cooperative society, Local Self-Government, Public Health, Medical Excise etc. to the 'provincial' sphere under the Devolution Rules and the expansion of the Legislature intensified the Secretariat organization too, which needed expansion. A reorganisation was undertaken in which the number of Secretaries was increased to six[7]. It was felt that a permanent officer of responsibility who knows the contents of all files as a repository of departments traditions and standards and who could deal with routine cases of all kinds and also could afford relief to his superior officer was needed. The Registrars, were, therefore, replaced by Assistant Secretaries. Due, however, to financial stringencies their reorganization was short lived. Important changes enforced were that the post of judicial Secretary was abolished. The Education Department of Secretariat was amalgamated with the office of the Director of Public Instruction (later known as Director, Education) and the Director was appointed as ex-officio Deputy Secretary, Education Department. Executive functions of the Board of Revenue were also transferred to the Secretariat[8]. The number of postings of Assistant Secretaries were also drastically reduced. A regrouping of departments into Branches, which took effect from the April, 1923, was made and except for the general Branch which was looked after by an Assistant Secretary under the direct control of the chief Secretary, other branches were known by the designation of their Secretaries as follows:

Branches	*Departments*
1. General	Appointments, General Political & Police
2. Revenue & Judicial	Revenue Scarcity, Legislative & Govt. Press
3. Education & Industries Education,	Industries, Agriculture, Civil Veterinary & Cooperative Society
4. Local Self-Govt. and Public Health	Medical, Lunatic Asylum, Excise, Income Tax, Custom Salt and Stamps
5. Finance	Finance, opium Excise, Income Tax, Custom Salt and Stamps
6. Executive	Secretariat, Establishment and Accounts Library, Record Room Issues section and General Executive duties

A revised "Secretariat Manual", which contained comprehensive instruction on office organization, disciplines and office procedure was brought out in 1925[9].

The Governor had since then made Lucknow his residence more because the Legislative Council was established here in 1921. The Department of Secretariat had also started shifting to Lucknow gradually and on the recommendation of the PannaLal - Macleod Committee appointed in 1932 to suggest ways and means of securing efficiency and economy in expenditure on the Secretariat, the exodus was completed by 1935. The Institution of "Camped Office" also came to close. The Irrigation and Building and Roads Branches of the Public Works Department continued to be separate entitle and as finance was a reserved subject under the Reforms, it was also made a separate unit in the Secretariat. Consequent on the changes introduced by the Govt. of India Act. 1935, in the shape of gradual devotion of administration from the Govt. of India to the State Govt., there was further expansion of the Rules of Executive Business were framed in 1937 for the transfer of business under the new constitution. Numerous new departments were added and existing departments were split many sections, thereby increasing the number of Secretaries, other offices and staff to handle the increased responsibility.

On December 5, 1938 these parts of the Irrigation and Buildings and Roads Branches which functioned as the Secretariat were amalgamated into one office, called the Public Works Secretariat under a non-technical Secretary to Government and the Chief Engineers were left with function of Heads of Departments only. The Secretariat thus, consisted of three distinct units: Finance Secretariat, Public Works Secretariat and the rest known as General Secretariat.

The establishment of the non-gazetted ministerial staff in the General Secretariat was further decentralized and placed under the control of various secretaries in 1951 on the recommendations contained in the Gopal Krishnan Report of 1950. Nevertheless the anachronism of keeping the three units as separate entities under the changed constitution continued until 1957 when these were merged into one composite unit.

After independence, the State took on itself the paramount task of the development of its economic resources. The normal growth of the Secretariat and the field department took place on this account[10]. The Secretariat work was distributed amongst the following 48 Departments which, except for a few, are sub-divided into two or more sections[11].

Appointment

Confidential

Vigilance

Civil defence

Cultural affairs and Scientific research

General administration (Recoganisation)

General administration

Language

Secretariat Administration

Petition

Finance

Public Work

Irrigation

Power

Agriculture

Animal Husbandry and Fisheries

Forests

Co-operation and Cane Department

Home (Police)

Home (Jails)

Home (General)

Judicial

Legislative

Election

Revenue

1. Revenue (Scarcity)
2. Education
3. Technical Education
4. Harijan Sahayak
5. Social Welfare
6. Municipal
7. Local Self-Government
8. Housing
9. Industries
10. Excise
11. Information
12. Relief and Rehabilitation
13. Panchayati Raj
14. Community Development
15. Planning
16. Labour
17. Transport
18. Medical
19. Public Health
20. Land and Civil Services
21. Rent Control
22. Border Development
23. National Integration (since July 1968)

The change described above have effected in the foregoing setup of the Secretariat during the last couple of decades in the State. The current position with regard to the number and the organization of the Secretariat is discussed in the paragraph below.

The initial decades after independence could not witness

much change in the organizational structure of the Secretariat of the State. However, generally as happens with all big organizations, in its journey over time especially during the last 2-3 decades, there has been sea change in the composition and functioning the U.P. Secretariat. The number of departments reported earlier as 48 has now increased to 74 i.e. an increase of about 54 per cent during a couple of decades. This tremendous increase in the size of the Secretariat is reflected in its strength of officers belonging to different services. The following figures show the proliferation of officers in U.P. Secretariat.

PROLIFERATION OF OFFICERS IN U.P. SECRETARIAT

Principal Secretary		*Secretary / Sp. Secretary*		*Joint Secretary / Under Secretary*	
NIL	33	21	158	114	259
---	---	---	---	---	---
---	---	---	---	---	---
1966	1998	1965	1998	1965	1998

A look at the figures given above shows the quantum of increase in the number of officers at the U.P. Secretariat. The increase has been tremendous keeping in view the short span within which this has been allowed to happen[12]. As organization grows in size, these often become unmanageable or at least difficult to manage. More often than not they remain mostly engaged in either house keeping function or day to day problems that demand quick reacting responses. The basic function of broad policy formulation and action get often shifted to back burner. The more important tasks get delayed or neglected unless innovative steps are taken to keep them on the right track. Apparently the U.P. Secretariat has grown enormously in size without a design, both vertically and horizontally, losing its vitality, purposes and elan the process. The rules, regulations and Government order prescribing Secretariat procedures and distribution of powers among different structures, level and departments have been periodically revised but the existing procedures are now in many ways impediment. This has to be rationally looked into for better functioning and efficiency in governance.

The Secretariat is the hub of the governmental activity in the State. Despite its strategic importance, a widely shared view is that for variety of reasons, its functioning has degenerated to an extent that presently it seems to be doing nothing more than fire-fighting in a very ad-hoc manner on a day to day basis, instead of the evolving and implementation of policies, programmes, procedure and laws according to a long-term perspective. It is not adequately realized that the functioning of the Secretariat department has far reaching impact on the over all working of the State Government. Some senior officers have told that the Secretariat working has all but collapsed. This is a view one may not share but it speaks of the widespread concern for the deterioration that has set in. The various factors affecting the decision making in the Secretariat is being taken up in another chapter of the study report.

In the very scheme of things the task of the Government at its headquarter, located in the Secretariat, should be formulating policies for accelerated development, for peace and tranquillity and in regard to laws, rules and procedures and working out the modalities for translating policy decision and legislation into implementation programme. One can argue out a number of genuine governmental functions which the Secretariat ought to perform. However, the general impression is that functions other than those relating to establishment matters have presently taken a back seat and that most of the Secretariat time is consumed in dealing with subsidiary matters relating to transfer, public grievance petitions etc. which should normally be handled and disposed of at lower levels in the district itself.

Proliferation in the number of departments, offices and staff, more specially officers, has made the Secretariat flabby.

Over 40% of the total number of available IAS officers in the State (excluding those on Central deputation) are deployed in the Secretariat making the organization top heavy. Location of Secretariat department in a number of building, with the officers and the attached office housed in separate premises has further worsened the working. The proliferation of staff has already adversely affected the working conditions as well as the work culture[13].

This is a brief description of the Secretariat, the structure and working. The other aspects dealing with decision making process and governance are taken up later[14].

PROFILE OF THE RESPONDENTS

Information about the socio-economic profile of the respondents was collected through a structured questionnaire. The information so received is being presented in the following paragraphs.

SERVICE CADRE

The respondents were asked to mention the service they belonged to. The information so received shows that, about 29 per cent respondents belonged to the Indian Administrative Service whereas 26 per cent were from Provincial Civil Service cadre. However, owing to larger number of employees in the Secretariat belonging to Subordinate service, their percentage in the sample has also gone up to 45 per cent.

Sex Ratio

The distribution of respondents according to sex shows that, majority of the respondents were male and only 9 per cent were females.

Age Group

Agewise distribution of respondents shows that the percentage of respondents upto the age group upto 35 years has been low i.e. about five per cent only. The maximum percentage of respondents belonged to the age group of 36 and 44 years is 41.38 per cent. The next higher percentage i.e. 36.21 was in the age group of 45 and 52 years. The percentage of respondents in the age group above 52 years of age was about 17.24 per cent. It shows that the respondents belonged to different age groups which would be reflected in the response pattern as well.

EDUCATIONAL QUALIFICATION

Information was also collected about the educational qualification of the respondents. The distribution of respondents according to the educational qualification was categorized only in three broad heads viz. graduate, post-graduate and professional. As could be seen from the figures given in the table majority of the respondents (51.72) had post graduate degree whereas a limited percentage 5.17% respondents had a graduate degree only. It is interesting to note that a considerable percentage of respondents (43.11%) had professional degree of one kind or the other.

LENGTH OF PRESENT POSTING

Information about the length of present posting was also collected from the respondents to know as to how long they have been on the present job. The information shows that a considerable percentage of respondents have had their length of service on the present posting as only one year. However, the higher percentage is reported (36.21%) as 'more than three years' by the respondents. This is because of the fact that the considerable percentage of respondents belonged to subordinate class since whose tenure in the secretariat is rather long. Owing to this the percentage of the respondents in 'more than three years' response category is also higher. The percentage of respondents in the case of 1-2 years is the smallest.

CASTE COMPOSITION

Information about the caste composition of the

respondents was also collected. The information shows that maximum percentage of respondents belonged to general category. However the sample did have the representation of other sections of the society as well including the Scheduled Caste and Other Backward Caste comprising 8.62 and 13.75 per cent respectively. The percentage distribution of respondents may not be in proportion to the population percentage, but it did provide us a fair representation of opinion about the subject of the study.

REFERENCES

1. Zaheer and Gupta, *Organization of the Governor of U.P.* S. Chand and Co. New Delhi, 1970 p. 40.
2. *Ibid.*
3. *Ibid.*
4. *Ibid.*
5. It was again bifurcated in 1970.
6. The old order of separate office was restored in 1939-40 and the Director of Public Instruction ceased to be ex-officio Dy. Secretary April 1939.
7. The Secretariat Manual was last revised in 1953. Most of its provision now need change and it is under revision.
8. *Ibid.*
9. *Ibid.*
10. T.N. Dhar, Secretary, IIPA, U.P. Regional Branch, of U.P. 1999, Lucknow
11. *Ibid.*
12. *Ibid.*
13. *Ibid.*
14. *Ibid.*

Chapter 3

Governance: Decision-Making—Value Frame

The governance could be termed as a soul in the entire body-politic. It is the quality and the essence of governance that would generally determine the nature and the quality of the state system. The concept of governance as such may be as old as the state system toady. It however, gains added importance in democratic system. One is reminded of Alexander Pope's observation that 'let the tools contest for the form of government, that government is best which is administered best'. It is a commentary on the quality of governance or it would determine the administrative process and the consequential attainment. Governance is authoritative allocation of values by those in power through decision-making. Since governance is the process of decision-making and the implementation of those decision, governance focuses on the actors' who make decision and the structure that are involved in the decision-making[1]. The governance, in fact, is given shape by those who mann the formal structures, the process involved and the behavioral parameters adopted.

The Constitution of India sets the broad parameters of governance for modern India which is designed to be a democratic, socialist, and a 'secular' Republic[2]. By implication the modern India also setforth the welfarist approach for future governance and growth. Over the years, however, there has been a question-mark on the quality of governance.

Irrespective of the political party in power, the quality of governance has remained questionable. The decision-making process and the quality of decision often determine the level of governance. In the present part of the book we are trying to analyze the decision-making as a process and the various variables influencing this process with a view to oversee the question of governance.

Decision making is an integral part of the management of any organization. The entire administration, its efficiency and effectiveness much depends upon decision making, the act and the process. Decision making have several facets too. It is an activity or action based on calculation, reasoning and purpose. An act on activity resulting from spontaneous behaviours, manifesting reaction and sudden emotional impulses could not be considered decision and the process as decision making. Simon has propounded his paradigm of decision making in his Administrative Behaviour providing an orientation of rationality. The impact for decision making, deliberate or otherwise can cause damage to the system.

Administration is ordinarily discussed as the art of 'getting things done'. Emphasis is placed upon process and methods for ensuring incisive action. Principles are suitable for securing concerted action from groups of men. In their discussion, however, not very much attention is paid to the choice which prepares all action to the determining or what is to be done rather than to actual doing. Although any practical activity involves 'both' 'deciding' and 'doing', it has not commonly been recognized that the theory of administration concerned with the process of decision as well as the process of action. This neglect perhaps stems from the notion that decision making is confined to the formulation of overall policy. On the contrary, the process of decisions does not come to an end when the general purpose of an organization has been determined. The task of 'deciding' pervades the entire administrative organization quite as much as does the task of doing indeed it is integrally tied with the latter. A general theory of administration has to include the principles of organization that will ensure correct decision making, just as it must include principles that will ensure effective action. In the discussion about governance in the present part, we shall broadly touch upon the themes of decision making and the subsequte the question of governance and good governance are addressed in the next part of the book.

Driven home by this empirical observation, Simon mentions that decision making includes on the main function of an administrative organization transcending all levels and modes of administration. He was naturally led to advocate that the theory of administration should treat decision making as the plank of its framework adopting the organization to the latest technical requirements[3]. Simon's analogy of decision making covered largely the psychological, sociological and political problem of every form of administrative organization. As a matter of fact, his own essays about decision making would not have some rich tributes but for their dependence on inter-disciplinary moorings.

In discussing the subject of decision making it is pertinent to focus on one or more of the three things: (a) the decision making process, (b) the decision maker, or (c) the decision itself[4]. It is interesting to note the virtual absence of a definition for the term decision. To say that a person has made a decision may mean that he has started series of behavioural reaction in favour of something, or it may mean that he has made up his mind to do certain action. But the most common use of the term is this: to make a judgement regarding what one ought to do in a certain situation after having deliberated on some alternative courses of action. Simon, the main contributor in decision making thought says that decision making comprises of three principal phases: finding occasion for making decision, finding possible courses of action and choosing among courses of action[5]. Others have also tried to provide some thoughts on decision making. However, as an academic discipline it is still relatively very young. It is only since the Second World War that operation research, statistical analysis and computer programming have imparted 'scientific' aura the process of choice only within the last about two decades that the behavioural sciences have begun to contribute to this body of knowledge.

SCOPE OF DECISION MAKING

Decision making can occur at several levels. The first and perhaps most basic level is that of the individual acting to satisfy his basic needs. According to Maslow, human beings are motivated by a hierarchy of needs. Different kind of needs have different levels of motivation linking the hierarchy of needs with motivation. The concept of hierarchy of human needs provides useful framework for analyzing individual decision. Much of the decision making accomplished by an individual relates to the solution of problem-personal, employment or social problem.

In complex social situations individuals usually find need satisfaction as members of groups that have particular purposes. Often they must compromise their personal decision if the group is to arrive at a consensus. Therefore, group represents more than just a collection of the desires of the individual members. The scope of decision making does not stop at the level of the group. A decision made by a particular group may commit the organization to a given course of action. More likely such decision will be made by several groups, be reviewed at several levels in the formal structures and centrally be ratified by the chief executives. A decision at the level of the organization is, therefore, much more complex than a decision made at the level of the group.

The scope of the decision making further extends beyond the managerial process at the level of the organization. Decisions are also made at the level of the total society. The scope of decision making is indeed wide. It commences at the level of the individual and extends to the deliberation of the groups that comprise the organization.

TYPOLOGY OF DECISION

Various experts in the field of decision making theory

have advanced numerous ways of classifying decisions. Perhaps the best known of these classifications is the distinction proposed by Simon between programmed and unprogrammed decision. According to him the decisions are programmed to the extent that they are repetitive and routine to the extent that a definite procedure has been worked out for handling them if a particular problem recurs often enough a routine procedure will usually be worked out for solving it.

Decisions are non-programmed to the extent that they are novel, unstructured and consequential. There is no cut and dried method for handling the problem because it has not arisen before or because its precise nature and structure are elusive or complex or because it is so important that it deserves a custom tailored treatment[6]. Drucker made essentially the same distinction as Simon, but he labeled programme decision 'generic' and unprogrammed decision 'unique'[7].

Thompson imparted a dynamic quality to the foregoing classification of decisions with his two dimensional scheme for decision making strategies. In this scheme there are the following four basic strategies:

(a) Computational Strategy: there is considerable certainty regarding cause / effect relationship regarding possible outcomes.

(b) Judgmental strategy: Preferences regarding possible outcomes are strong but cause / effect relationships are highly uncertain.

(c) Compromise Strategy: there is a good deal of certainty regarding cause / effect relationship, but preference for possible outcomes are not strong.

(d) Inspirational Strategy: Reference for possible outcomes are not strong and there is considerable uncertainty regarding cause / effect relationships[8].

A closer analysis would suggest that it is not too difficult to detect a higher degree of commonality among these several classification schemes for decision and decision making strategies. Essentially each scheme can be reduced into two basic categories involving the matched pairs of (i) routine and non-routine, (ii) recurring and non-recurring, and (iii) certainty and uncertainty.

Decision theory is still a new field. Most current orientation have strong quantitative emphasis that focuses on the decision itself rather than on the process within which the choice takes place. It is significant that a decision is simply a means to an end, and not, as it frequently assumed, an end in itself. The end is, of course, the out come that will result from the choice or outcome that hopefully will attain the objectives that gave rise to the integrated process of decision making[9].

Decision making is a dynamic process rather than a fixed procedure or a choice made at a particular point of time. The decision making process itself is compact of several sub processes. The dynamics of the decision making process is fairly forward although some complexity may arise in the several sub processes.

RATIONALITY IN DECISION MAKING

Decision making process is also surrounded by a debate about the rational and non-rational decision making. The term rational decision making epitomizes the general confusion and varying interpretation surrounding this most important activity in the formal organization. On the one hand, the term is often applied solely to the choice itself, that is, the decision is rational or non-rational. On the other hand, it is frequently used to describe the behaviours of the decision maker as he proceeds towards the selection of a preferred alternatives.

The concept of rationality is subject to widely varying interpretations. Rationality is used frequently in connection with the overall process of making decision or some part of the behavioural pattern comprising that process. An action is judged to be rational or non-rational according to the perspectives of the participants or the observers. Occasionally, the decision itself as well as the process of making it is subjected to a similar judgement.

According to Simon, Rationality is concerned with the reduction of preferred behaviour alternatives in terms of some system of values whereby the consequences of behaviour can be evaluated[10]. Simon also classifies behaviour in the formal organization as objectively, subjectively, deliberately, organizationally or personally rational, according to the values or goals of the individual[11]. Most individuals in making decisions believe that behaviour is completely rational or logical at the moment. There is an objective and several alternatives from which a choice is made to achieve the objectives.

Rational behaviour in the decision making process simply involves the evaluation and selection of some relevant alternatives based upon a perceived advantage accruing to the decision maker. The advantage that is intended to satisfy the objectives may be rational or psychological, and it may or may not be attained. All that is necessary is to make the choice a rational one is the existence of an objective and the selection of some alternatives that promises to meet the objectives as seen by the decision maker[12].

This is not to suggest that the choice will not involve analysis or calculation. Further, the choice may be made under conditions of perfect information or complete uncertainty and still be rational. More likely it will be made with incomplete or imperfect information and therefore, involves some amount of risk.

FACT AND VALUE IN DECISION MAKING

Fact and value paradigm in decision making has particular significance. Simon illustrates the harmonious blending of value and fact as the structural requirements of decision making. He says that "in so far as decision lead towards the selection of final goals, they will be called value judgments" so far as they involve the implementation of such goals they will be called "factual judgments[13]. It is generally observed that goals are final objectives of a public organization or governmental organization are usually spelled out in a very abstract and general terms such as justice, general welfare, and liberty. Even value and fact elements are combined in some cases in a single objective. This discussion should, however, not give a notion that purposiveness is another name for rationality. Simon distinguished the two in a subtle but vividly clear fashion when be says that behaviour is purposive in so far as it selects alternatives which are conducive to the achievements of the previously selected goals[14]. It is, thus, obvious that all decisions are subject to some element of compromise and this reflects the regulation elements in decision making.

Decision making process naturally involves question of **fact** and **value**. Simon tried to explain the fact and value paradigm and said that decisions are something more than factual proportions. To be sure, they are descriptive of a future state of affairs, and this description can be true or false in a strictly empirical sense, but they posses in addition an imperative quality - they select one future state of affairs in preference to another and direct behaviour towards the chosen alternatives. In short, they have an ethical as well as a factual content[15]. Simon thus argues that decisions could be both fact and value oriented.

In administration, the mixed character of the ethical is usually fairly obvious. A municipal department may take as objective the providing of recreation to the city's inhabitants. Their aim is thus further analyzed as means towards building bodies, using leisure time constructively, preventing juvenile delinquency and a host of others, until the chain of means and end is traced into a vague realism labeled 'the good life'.

MEANS AND ENDS

The fact and value premises are related to means and ends respectively. In the process of decision, those alternatives are chosen which are considered to be appropriate means for reaching desired ends. Thus, rationality, has to do with the construct of means – ends chain or their kind. While elaborating the environment of decision making, the decision about means and ends is again taken up. In this context, Simon puts-forth certain caution against the possibility of leading to inaccurate conclusions through the analysis of rational behaviour in terms of means-ends hierarchy.

There are three cautions which Simon has referred which are discussed here. Firstly, rational decision making always requires the comparison of alternative means in terms of respective ends to which they will lead. This means that efficiency – the allotment of maximum value with 'limited means' must be a guiding intention in administrative decision. Secondly, in actual situation, a complete separation of means from ends is usually impossible for the alternative means are not usually valuationally natured. Thirdly, the means end terminology items as enunciated by Simon tend to observe the role of time element in decision making.

In fact the goals may be dependent for their force on other more distant ends leads to the arrangements of these goals in hierarchy in each level to be considered as an end relation to the level below it and as means relative to the level above it. Through the hierarchical structure of ends, behaviour attain integration and consistency for each member of a set of behaviour alternatives is then weighed in terms of a comprehensive scale of values --the 'ultimate' ends. In actual behaviour a high degree of conscious integration is seldom attained[16].

The hierarchy of means and ends is as characteristic of the behaviour of organization as it is of individuals. It is also as true of organizational as of individual behaviour that the means-ends hierarchy is seldom as integrated, completely connected chain. Sometimes the lack of integration in an organization in mean-end hierarchy is due to refusal of the policy making body to decide 'lot' issue of policy[17]. Both organization and individuals, then fail to attain a complete integration of their behaviour through consideration of these means - ends relationship.

It is clear that the 'means-end' distinction does not correspond to the distinction between fact and value. What then is the connection between the two sets of terms? Simon says that connect a value with the situation releasing it, and these situations, in turn with the behaviour that produce them. Any element in this chain may be either 'mean' or 'end', depending upon whether connection with the value and of the chain, or its connection with the behaviour and of chain, is in question[18]. Simon further explains that the means character of an element in a means-end chain will predominate if the element is towards the behaviour end of the chain; the end character will predominate in the element is descriptive of the consequences of behaviour[19].

So far we have tried to present a discussion about the relationality element in decision making along with the concept of fact and value followed by means-end question. Attempt has thus been made to explain the concept of decision making and the factors taken into account in the decision making process. It is now proposed to take up the environment of decision making process so as to understand the compulsions on individuals who operate in an environmental setting.

THE ENVIRONMENT OF DECISION MAKING

Organizations do not exist or operate in vacuum. They are part of a larger aggregate such as the economic system, the social system and the political system of the nation / state. Therefore the decision makers must take account of the inputs of the environment in arriving at a decision, and they must consider the effects of that choice on the manifold forces comprising the environment. In particular, the satisfying decision makers in the open decision model must be keenly aware of many uncomforts external variables that shape his objectives and aspirations, channel his search, limit his range of limited alternative and then hamper his selection or the one alternative that is good enough for his intended purpose. To transform an intellectual state into an operational reality, however, a choice must be implemented effectively. The effectiveness of the implemented decision is conditioned considerably by the environment[20]. Any decision, therefore, gets influenced and conditioned by the environment about which the decision makers ought to be conscious of while choosing the alternatives.

THE ENVIRONMENT AS A SYSTEM

It is useful to conceive of the environment as an open system within which the satisfying decision maker in a formal organization functions as an essential element, receiving inputs on the one hand and dispensing processed out puts on the other. The system inputs consist of information, energy and materials, and the system outputs are the satisfying decision[21].

The objectives that underline the decision making process in the open system may be formulated at any level, they are ratified at the organizational level. Most likely they are the products of a multiplicity of interacting environmental forces. The establishment of objectives triggers the search, and the environment is scanned to uncover relevant alternatives from which a satisfying choice can be made. The search results in a stream of environmental inputs which undergo an intensive process of analysing and evaluation to determine the degree of compatibility with a fulfillment of the objectyives[22]. This process continues till the decision maker feels that within given constraints the most satisfying alternatives has been uncovered at which times a choice is made and implemented.

We have earlier referred to about three types of systems namely , (a) the economic, (b) the social and (c) the political system. We shall briefly focus upon these systems which work as an environmental force.

(a) The Economic System

One of the basic environmental forces affecting the

decision maker in a formal organization, is the economic system. The economic system include employees, customers, competitors, other industries and the operation of the federal government is regulating business activities to protect the public interest and is using monetary and fiscal policy to influence the availability by money and the cost of the capital. The economic system may be conceived of as a market place where a seller and buyer interact to established the prevailing price structure for a wide range of commodities and services.[23]

One primary function of the economic system is the allocation of scarce resources i.e. lands, capital and management. The economic system provide some value measurement device, such as prices, wages, or quotas to apportion these resources among competing users in a control of basic supply- demand relations. This function is explained in the context of a capitalist economic system. But it is partly applicable to other economic systems too wherein controls become necessary in public interest.

A second function of the economic system is the maintenance of the effective aggregate demand for the products and services of that system. There must be an adequate supply of money and a desire on the part of the general public to spend that money. A third function of the economic system is to provide for the timely and efficient distribution and exchange of the goods and services. The transfer of goods and services from producers and sellers to consumers and byres is a principal determinant in the effective functioning of the economic system.

(b) Social System

Another of the several environmental forces that exert a significant impact on the decision making process in formal organization is the social system. Persons and Shil define the social system as:

- A social system is system of the action of individuals, the principal units of which are roles and constellations of the roles. It is a system of

> differentiated actions, organized into a system of differential roles. Internal differentiations which is a fundamental property of all systems requires integration. It is condition of the existence of the system that the differentiated roles must be coordinated either negatively in the sense of the avoidance of disruptive interference with each other or positively in the sense of contributing to the realization of certain shared collection goals through collaborated activity[24].

A comprehensive way to view social system is through both its structure and performance. The structures relate to the matter or mass of the system that occupies the space. The performance relates to the expeditors of energy of the action that takes place within the system. At any level, structure and performance are aspects of the several processes which, taken together, comprise the social system.

Specific social problem confront organizational decision making at every turn, for example, the current emphasis on equal rights for women and minority group is impossible to ignore. The increasing pre-occupation with the pollution of air and water is a primary consideration in the operation of the organization. The strategies with which to combat these variables in the social system are of increasing concern to organizational decision makers[25]. The problem arising out of social changes too have to be addressed by the decision makers in the decision making process.

(c) The Political System

A major environmental force which decision-makers in the formal organization must contend is the political system. This system is perhaps even more subtle in its effects than the economic system which is characterized by definable institution, process and markets. The political system is founded on the concept of governance, which is the exercise of an authority that seeks not merely the right to direct, and to lead but to control.

The governments at all levels wield the power to shape, to a larger degree, the environment in which the citizens live. It is, therefore, a major determinant in the environment of organizational decision making. The influence of the government is felt by organization namely in the form of law and regulations pertaining to the establishment of a legal entity and the subsequent conduct of its regular operation. In the primary instrumentality of the political system, the governance personally expresses the will of the people, and the more complex the operation of the organization become the more power is granted to the government to protect the public interest. The organizational decision maker is not only influenced in his choices by the government but also by the real makers of the public policy, i.e. the various interest groups that comprise the locus of the real political power in a pluralistic political system. For example, intellectuals want organization to behave in a more socially responsible way, minority groups want organization to make special provisions for their needs and labour leaders want an increasingly larger share of productivity gain as well as a larger voice in the management of the organization[26]. In an undeveloped social process, political system further gets influenced by other groups as well which influence the decision making.

VALUES FOR DECISION MAKING

The personal values of the decision maker and the values of the organization of which he is a part, exert an influence on the process of choice. In the setting up of objectives, the values of the organization must be considered by the decision maker. The search will reflect the personal values of decision maker as he attempts to relate the alternatives uncovered to the organizational objectives. Once the search is completed, the personal values of the decision maker conditioned by the values implicit in the organizational objectives influence the comparison and evaluation of alternatives. With the making of the choice, the implementation phase is initiated and in the process of executing the decision, the values of the organization usually take precedence over those of the decision maker.

Finally, in the follow up and control phase of the decision making process the organizational values are inflicted in the standards that provide the basis for measurement and corrective action[27].

At all points in the integrated process of decision making, the personal values of decision maker may conflict with the values of the organization. In the process of such conflicts, the decision maker must frequently accommodate and subordinate the personal values to the organizational objectives. If this is not allowed to happen, in that case, the organizational objectives might suffer[28]. On the one hand, the process values of the decision maker are considered important and realistic, on the other hand, their subordinates in case of conflict with organizational objective is also perceived as realistic in the decision making process.

REFERENCES

1. Barthwal, C.P., *Good Governance in India*, Deep and Deep Publication Pvt. Limited, New Delhi, 2003, p. 5.
2. *The Constitution of India*, The Preamble.
3. Simon, *Administrative Behaviour*.
4. E. Frank Harrison: *The Managerial Decision Making Process*, Honghton Miffin Company, Boston, 1975, p. 3.
5. Herbert Simon, *The New Science of Management Decision*, New York, Mc Graw Hill, 1970, p. 31.
6. *Ibid*. pp. 5-6.
7. Peter F. Drucker: *The Effective Executive*, Harper and Row, Mc Graw Hill, New York, 1970, p. 31.
8. James D. Thompson, *Organization in Action*, McGraw Hill, New York, 1967, pp. 134-135.
9. S. Frank Harrison: *The Managerial Decision Making Process*, Houghton, Miffin Company, Boston, 1975, p. 17.
10. Simon, *op. cit*., p. 75.
11. *Ibid*. pp. 76-77.
12. E. Frank Harrison, *op. cit*., p. 61.

13. Herbert Simon, *Administrative Behaviour.*
14. *Ibid.*
15. *Ibid.*
16. *Ibid.*, p. 63.
17. *Ibid.*, pp. 63-64.
18. *Ibid.* p. 74.
19. *Ibid.*
20. E. Frank Harison, *op. cit.*, p. 85.
21. *Ibid.*
22. *Ibid.*, p. 87.
23. *Ibid.*, p. 91.
24. Talcott Parsons And Edward A Shills (Eds.): *Toward a General Theory of Action*, Harper and Row, New Yark, 1917, p. 197.
25. E. Frank, *op. cit.*, p. 103.
26. Richard Eeb and Clavence Walton: *Conceptual Foundation of Business*, Rev. Ed., Homnod III Richard D. Irmin, 1969 p. 405.
27. Hodge and Johnson: *Management and Organizational Behaviour*, p. 73.
28. E. Frank Harrison, *op. cit.* pp. 99-100.

Chapter 4

Good Governance: Concept and Context

After having briefly analysed the issue of governance, it is now proposed to examine the issue of good governance, concept and the context. It may be argued that the governance by implication has to be good otherwise, it cannot survive for long. In this context one is reminded of Aristotle's famous dictum that 'the state came into existence for life but continues to exist for good life'. The state and the government would cease to exist in the absence of good governance, if they fail to provide for a good life for the citizenry. The focus of the present work attempts to look at the question of good governance from that view point wherein it would try to see if there is any inter-linkage between caste and governance. This part of the book attempts to briefly discuss the concept and the context of good governance.

THE CONCEPT

The theme of good governance has encouraged considerable debate world over. More specifically, since the 1990s, the term has acquired currency in contemporary world scenario where the World Bank entered into the scene. It has been suggested that the use of the term 'good governance' is in the context of western aid to developing societies and the benchmarks fixed by the former on the latter as a condition for giving that aid. We would, therefore, confine our discussion on good governance in the contemporary context and leave the historical interpretation as that may not help in the present debate.

It is argued that the World Bank in one of its document in 1989 highlighted the concept of good governance in the context of Sub-Saharan Africa. It highlighted the concept of good governance and identified the following four dimensions-

(a) Public sector management

(b) Accountability

(c) Legal framework for development; and

(d) Information and transparency

Further, in the year 1992 the World Bank, document entitled **Governance and Development** mentioned that 'Good Governance is central to creating and sustaining an environment which fosters strong and equitable development and it is an essential component to sound economic policies'. In the process the document identified three aspects of governance:

(a) the form of political regime;

(b) the process by which authority is exercised in the management of countries economic and social resources; and

(c) the capacity of governments to design, formulate and implement policies, and in general, the discharge government function.

The World Bank has, however, identified the following elements in good governance:

(i) Political accountability

(ii) Regular election to legitimize the exercise of political power.

(iii) Participation of various social, economic, cultural and professional groups in the process of governance.

(iv) Rule of Law.

(v) Independence of Judiciary.

(vi) Bureaucratic accountability

(vii) Freedom of information

(viii) Transparency

(ix) Efficient and effective administrative system.

(x) Cooperation between government and civil society.

Apart from the World Bank document several other international documents have tried to identify indicators for good governance wherein they tried to link the international aid to development and to good governance. The 1990s witnessed the emergence of some commonly coined indicators for measuring parameters of governance (as identified in UNDP, 1996 workshop), factors which could help foster better bonding between the government and the governed. They are:

- Participatory
- Sustainable
- Legitimate and acceptable to the people
- Transparent
- Promotes equity and equality
- Promote gender balance
- Tolerates and accept diverse perspective
- Develop resources of governance
- Strengthens indigenous mechanisms
- Operates by rule of law
- Efficient and effective use of resources
- Engenders and commands respect and trust
- Define and take ownership of national solutions
- Enabling and facilitation

- Service-oriented
- Accountable

The Organization for Economic Cooperation and Development (OECD), comprising of forty-eight aid giving countries, in its document attempted to link development assistance with (i) participatory development, (ii) human rights, and (iii) democratization.

It has been argued that good governance being value-oriented is a sub fiction concept. It has different meaning for different people[1]. It may be of interest here to mention some of the indicator for good governance provided by Kautilya which have relevance even in contemporary governance. They are as under:

- King as a servant of the state.
- Properly guided Public Servants.
- Code of Conduct for Rulers.
- Fixed Salaries and Allowances to the king and public servants.
- Maintenance of Law and Order.
- Preventive and Punitive Measures against Corrupt Officials.

Looking into the parameters provided by the World Bank along with other international bodies and that of Kautilys's indication, it is rather difficult to find a comprehensive definition of good governance, which could include all the parameters. At best we can say that good governance is associated with efficient and effective administration in a democratic framework, an administration considered to be citizen friendly, citizen – caring and responsive one.[2] Amid global efforts to forge some kind of consensus on the ramification of good governance, there are efforts to operationalise the concept within the South Asian region. For example, the Pakistan-based Mabbubul Haq Human Developmental Centre has recently come out with a

Human Governance Index (HGI) as "a composite index of indicators meaning economic, political and civic governance."[3]

It is also argued that there are several actors and as many view points in a given society. Good governance requires mediation between different interests of the society so as to reach a broad consensus in the best interests of the community. For better governance, it is important that the actors in the community develop a culture of partnership and a sense of personal commitment towards national goals. The goals of good governance can be achieved only by making the interpersonal boundaries more flexible and permeable. No society has progressed whose people are divided on caste, class or ethnic lines. Cultural pluralism should not act as a deterrent to social progress.[4]

It may be noted that the term 'good governance', as currently used in the Third World Countries, is not only of recent origin, but is also of the western variety. It is, therefore, no wonder that varying conceptions of good governance exist in these countries.[5] Despite its popularity and wide currency, the concept entails ambiguity, and a bewildering variety of meanings and conceptualizations. Simply put it means any thing to anybody depending on one's perception of what is good and for whom? Understandably, it is equated with a wide range of ideas not always consistent with one another. For some it may mean leas government and for some others it may mean more government[6]. It can however, be argued that there seems nothing unnatural when we observe variation in perception. But, inspite of the variation in perception, one can find a more acceptable definition and conception about good governance.

Without taking the discussion any further, it is now proposed to present the following indicators / parameters of good governance where one may claim some agreement. They could be:

1. Participation
2. Accountability

3. Predictability
4. Transparency
5. Rule of Law-Maintenance of Law and order
6. Freedom of Information and Expression.
7. Cooperation between Government and Civil Society organization

Let us briefy elaborate these parameters for a better understanding.

1. Participation

The focus in the good governance seems to be on a participatory approach in the matter of the state. It is the people who are at the centre of the governance. They take part in governance through formal institution of the State as well as through the non-governmental organization in the form of political parties, caste and religious groups NGOs, business groups. The people therefore, not only the agents of good governance through various means, but they are also the beneficiaries of good governance.

2. Accountability

Accountability is another core element in good governance. In the absence of accountability factor governance can be arbitrary and biased which can affect the public good adversely. With this in view the organization, public or otherwise, have to be accountable in one or the other way to make good governance effective and meaningful.

3. Predictability

The predictability factor is seen in the context of rule of law. It refers to: (i) the existence of laws, regulations and policies to regulate society; and (ii) their consistent application. Consequently there should be orderly existence of citizens and institutions in society. This would be possible only when the governance bases itself on the rule of law. It would then encompass both, well defined rights and duties, as well as mechanisms for enforcing them, and setting disputes in an impartial manner. Consisting of public policy also can be covered under predictability. It however, does not mean rigidity. Governments do need to respond flexibly to changing environmental factors.

4. Transparency

There has been a considerable focus on transparency factor in the governance these days countering the earlier 'wisdom' of secrecy in the governing process. There is greater awareness about the right of the people to know about things that are supposed to affect and influence their life system. Right to information thus, become a core component and it naturally asks for transparency in the governing process. It generally refers to the availability of information to the general public and clarity about government rules, regulations and decisions. It tends to eurb corruption.

5. Rule of Law: Maintenance of Law and Order

Maintenance of law and order is one of the core components in the concept of good governance because in its absence no state system can meaningfully survive. Maintenance of peace and tranquility has to be on the basis of the rule of law so that no discrimination could be attached on other grounds such a sex, caste, creed etc. According to Kautilya the chief duty of the king is to maintain law and order i.e. to protect the life, liberty and property of the citizens. The king will have to pay from his own pocket if he is found guilty of dereliction of duty on this score. As Kautilya suggested, 'Whatever the property of citizens robbed by thieves, the king cannot recover, shall be made good from his own pocket'.

Apart from the above reference of Kautilya, it was noted during period of Nawabs in Awadh, a Kotwal declared that there shall be no theft in the city, and in case if happens he promised the citizens to compensate the amount.[7] Such cases are a rarity, but none the less they pinpoint towards the importance of the maintenance of law order in governance.

6. Freedom of Information and Expression

Good governance being people – centered governance, it is in the fitness of the things that it provides for freedom of information and expression. It is one of the important parameter of good governance. In the absence of freedom of information, the traditional ways of secrecy and confidentiality would mark the governance which might hamper the concept of good governance it stands for.

7. Cooperation Between Government and Civil Society

Scholars have argued for cooperation between government and civil society for good governance. The discourse on 'civil society' in contemporary social sciences covers many areas of governance in democratic societies. The intellectual tradition of discoursing the nature of state and civil society is traced back to the contract theorists – Hobbes, Locke, and Rousseau on the one hand, and to Hegel, Kant, Marx and Engels on the other[8]. The State, according to Hegel, is not only responsible for the creation of civil society, but also for its sustenance for the purpose of serving a variety of collective interests. It is often argued that civility is the core of civil society. It is also equated with modernity, individual freedom, individuals moral autonomy, development or collective conscience and enlightenment, and above all, the societal self-organization, and people-centred notion of human well-being[9]. To operationalism these, there has to be a cooperation between the government and civil society. In the absence of this cooperation, good governance can hardly be achieved.

Taking an overall view, good governance should basically mean balanced governance. Government, private sector, non-government organization and cooperation should harmoniously share the responsibility of governance by protecting the interests and meeting the needs of the people[10]. It is also aimed at reducing the state intervention in so many areas apart from making an attempt to restructure the system to suit the requirements of good governments which is précisely people-focused.

THE CONTEXT

After having briefly discussed the concept of good governance it is now proposed to place the concept in the context, the context of India and more specifically the state of Uttar Pradesh, wherein we based our study. Since the focus of good governance by definition seems to be on the third world countries, India naturally formes the core in that discussion.

Let us have a brief historical review of good governance in the context of India before analyzing the same in the present context. Kautilya in his famous treaties Arthshastra has explained that – in the happiness of the subjects lies his happiness, in their welfare his welfare, whatever please him (personally) he shall not consider as good, but whatever makes his subjects happy, he shall consider good. This was the view, the advice, the comment on the relationship between the ruler and the ruled, the king and his subjects more than two thousand years back where the 'subject' remained at the core of governance. Different phases of Indian history tell us about different concerns in governance. In the present discussion one would like primarily to concentrate on the modern India, particularly the post-liberalization phase which is the phase of good governance.

It is being increasingly argued that civil society has a tremendous role in ensuring good governance in India. Civil society groupings in India do have the potential of focusing public attention on variety of problems and issues that confront the polity[11]. The role of civil society in good governance is now a familiar theme in India. An increasing role for non-governmental organization is prescribed to counter the ill-governance of the governmental organizations. But it is an unfortunateiity that in India we have large number of non-governmental organization that do not seem to be operating in tune with their people-serving objectives and in performing their advocacy role efficiently. Most of them are ill governed, mismanaged and corrupt as the government machinery. Only few of them have been able to live up to their objectives[12].

Governance in India has been a complex affair. India started its post-independent journey with the Congress as the only dominant party that pervaded the political scenario for nearly forty long years. The other political parties were of little significance in the governing process. With exception, there were relatively fewer distortion in thẹ Constitution. Besides, the socialistic policies followed by the Governments, both at the Centre and State, gave little scope to the rise of consumerism and that kept the needs of the people within reasonable limits[13]. Subsequent change in the political scenario of the country have made the governance a rather much complicated affair from various counts.

It may be noted that the Indian State is progressively withdrawing itself from many vital sectors of public governance in favour of private concerns, be it higher education, agricultural development, domestic industries or power supply. More and more profitable Public Sector Units (PSUs), such as BALCO, and BHEL are being privatized and many more are in the waiting list. There is a clear paradigm shift in so far as the impact of good governance in India is concerned. While the government is moving in tune with the requirements of globalised world, the actions have generated a crisis of governance by inducing unemployment, inequality and irresponsibility[14].

Whereas there may be presence of world bodies and otherwise too for good governance, but in India it has allegedly taken a back seat. In fact, most of the democratic governments of South Asia have strayed from the path of good governance, which is an integral part of democratic governance. Issues of common concern such as equality, liberty and justice have taken a back seat, while corruption, communalism and crimilisation have become the driving force of modern governments[15]. If good governance stand for transparency, accountability, openness and smooth information flow[16] or transparent and responsible application of authority[17] or it refers to "the creative potential of politics, especially with the ability of leaders to rise above the existing structures of the ordinary, to change the rule of the game and to inspire others to first take in efforts to move society forward in new productive directions", [18] then sadly most of these components are missing from the governmental method of working and its policies, programs and legislation.

The phase of liberalisation in India has also witnessed a new phase in Indian politics, the phase of coalition politics. Quite a few people in informed circles have started arguing that coalitions have come to stay in India. This coalition phase has further influenced the governing process. Basing the argument on the premise that coalition governments have come to stay in the contemporary political scenario, it would be out of place to mention that the present political scenario projects a crisis in governance[19]. We have witnessed the governance during almost six years of governance headed by Mr. Atal Behari Vajpayee the National Democratic Alliances (NDA). Severe criticism about the quality of governance was launched by the opposition with no avail. Ultimately the general election held in the year 2003 settled the issues when the NDA alliance was defeated, rejecting all the electoral forecasts about NDA victory. It was the verdict of the people on the quality of governance. As a matter of fact, the general mass of people were the victims of ill governance during that phase. This gave rise to a new phase of politics, a new alignment of political forces. The new coalition, the United Progressive Alliance (UPA), headed by the Mr. Manmohan Singh of Congress Party, supported by the left parties, is in the reign. How long their coalition will last is a matter of opinion and only time can tell. However, one thing is certain if this coalition fails to address the critical issues of governance, issues concerning the general mass of the people, this coalition is also destined to have the same fate as that of the NDA. It is, however early to make any rational observation about its future, as they have started only recently, but they are definitely feeling the pains of coalitional governance. It has been alleged that coalition governments have institutional politics of opportunism and corruption ... the explanation for emergence of immoral, corrupt, directionless and incoherent coalitional government lies in the logic of opportunistic politics which has become the hallmark of Indian political class[20]. The governance has thus, become a matter of convenience to legislative authority in which the 'good' becomes a causality.

There is no denying the fact that governance has to be growth oriented and for the public good. During the last five decades, government guided by the socialistic ideals of the Constitution has been the major player in the funding and execution of all major development programs and projects in the key areas of the economy. In this there has been some gains and shortcomings. However, during the last one decade deteriorating law and order situation, rising graph of crimes, criminalisation of politics and politicalisation of the administrative machinery, mega financial scams and scandals, falling standards, norms and values, all round have eroded public confidence in the government and the government machinery. Their promises and assurances carry little weight. Words of Government appear to have lost sanctity. There are more public official but little public service[21]. Under such a situation there is a big question mark on quality of governance.

The role of government is now largely becoming that of a 'facilitator' and 'regulator' of development. According to the Government of India's Economic Survey (2002-2003) the changed paradigm of the public sector now is on providing public goods and services without necessarily producing them. The Approach Paper to the Tenth five-year plan of the Government of India also suggest the need for such a shift in policy. It says, in many respects, development policy in future must take a break from the past. The government has over the years taken on itself too many responsibilities with the result that it not only marginalzed individual inclinations but also succeeded in imposing severe strains on its financial and administrative capability. The government has now realized, as it says the need to release latent emerging and stimulate private initiatives in various facets of our development process. Thus the new development policy highlighted in the Tenth Plan Approach Paper is to plan for an environment that provides ample opportunities for all to actualize their potential individually as also collectively for the nation as a whole[20]. This new endeavor is in a way a commentary on the status of good governance in India.

In the forgoing pages we have tried to put-forth the concept of good governance in the context of India. It is important to mention several opinions, and at times, even contradictory opinions have found place in the debate over good governance in India. The acid test for any government remains the fulfillment of the major and sub-major objectives to a satisfactory level. It could be in terms of transparency participation, maintenance of law and order, of growth perspective. In all these the Government, at different levels have to show near perfection in their achievement chart. But, suddenly, we have to go a long distance from here now in almost all the sectors of governance, leave aside the good governance, there may not be unanimity on the performance levels of various government at the centre and then at the state level owing to different perception and alliances, but the common Indian citizen had to react to these situation in his own way and he had shown that earlier on several occasions. The good governance, therefore, seem to be "distant, a far cry in India". It should not sound a pessimistic note as there is always a silver lining in the dark cloud.

REFERENCES

1. Subhash C Arora, "Concept of Good Governance" in *Good Governance in India,* C.P. Barthwal, Deep and Deep Publication Pvt. Ltd., New Delhi 2003, p. 33.
2. *Ibid.,* pp 33-37.

3. Anju Saran Upadhya, Issues Concerning Good Governance in Good Governance in India, C.P. Barthwal, Deep and Deep Publication Pvt. Ltd., New Delhi, 2003, p. 39.
4. D.P. Singh, Good Governance and Development, Indian Journal of Public Administration, New Delhi, Vol. L, No. 1, Jan-March 2004, p. 296.
5. S.P. Srivastava, Third World Development and Civil Society Contribution, in Good Governance in India, op. cit., p. 59.
6. *Ibid.*
7. S.N. Singh, Administration of Kingdom of Awah, Mittal Publication, New Delhi, 2004, p.
8. S.P. Srivastava, Third World Development and Civil Society Contribution, *op. cit.*, p. 64.
9. *Ibid.* pp. 64-65.
10. C.P. Barthwal, Good Governance in India, *op. cit.* p. 21.
11. S.P. Srivastava, Third World Development and Civil Society Contribution, *op. cit.*, p. 65.
12. *Ibid.* p. 66.
13. Sanjay Gupta, *India and the Crisis of Good Governance*, Good Governance in India, *op. cit.*, pp. 156-157.
14. *Ibid.*, p. 157.
15. *Ibid.*, p. 159.
16. *World Bank Report*, 1992.
17. NEDA – Netherlands Development Assistance.
18. Goran Hayden, 'Governance and Sustainable Livelihoods: challenging and opportunities', 1998, unpublished paper.
19. K. Kishore, Coalition, Administration and Society, in *Public Administration and the Coalition in India*, S.N. Singh, Bharat Book Centre, Lucknow, 2003, p. 47.
20. C.P. Bhambhri, Crisis of Coalition, *The Pioneer*, Lucknow, August 29, 2001.
21. Indian Journal of Public Administration, Vol. L, No. 1, Jan.-March 2004, dictorial, I.I.P.A., New Delhi.
22. *Ibid.*

Chapter 5
Caste: A Variable in Governance

Governance has become complex as several factors have a bearing on this. Since governance largely depends on the decision making process in a given environment, it also gets influenced by several factors. We have tried to analyse the question of 'fact' and 'value' in the decision making as well as the governing process in the earlier part of the book. The question of 'rationality' too has been attempted earlier. Some space has also been provided to the question of 'environment' in the decision making wherein social, economic and political environment has been discussed. It is now proposed to take up the issue of caste vis-à-vis governance in the context of U.P. We have tried to examine various factors which have a bearing on the decision making including caste which influences the governance. The present assessment is based on the primary data collected through a structured schedule (copy annexed as Annexure I) and information gathered through other formal and informal sources. The observation, therefore, are limited to the U.P. Secretariat, located at Lucknow and some field offices here.

The main centre of decision making of the Government of Uttar Pradesh is the Secretariat where the officers belonging to the Indian Administrative Service, Provincial Civil Service and Subordinate Service alongwith political masters are involved in the decision making. These officers belong to different socio-economic groups which allegedly has an impact on their working. Depending upon their socio-economic background, the response patterns too has varied. We have tried to collect information on various aspects of the service and their work. An analysis of the same is being presented in the following pages.

HEAD QUARTER – FIELD POSTING RESPONSE

The civil servants generally have to serve in the Secretariat head quarter as well as in the field areas during their service tenure. In order, therefore, to know as to which of this two postings, they enjoyed most, they were asked to give their opinion on this question. The responses received tell us that majority of the respondents (Table 5.1, Appendix II) about 55 percent enjoyed their field posting. Also considerable percentage of respondents (43 per cent) mentioned that they enjoyed the secretarial posting. The field posting option by the majority of the respondents may be generally because of the fact that in the field, the officers enjoy greater benefits and perks. There is greater scope for discretion which can be put to use either way. The officers come in close contact with the people and the interaction ought to be more in the field which has its own advantages. In view of these factors, the officers expressed that they enjoyed the field posting more as compared to secretariat.

In a subsequent question as to which place the working was difficult the respondents gave their opinion. The figures show that majority of the respondents (Table 5.2: 55 per cent) have expressed that working in the field was a difficult experience. The next higher percentage (31.04 per cent) has reported that working in the Secretariat has been difficult. It can, thus, be argued that working has been difficult at both the places, it is only a question of degree.

THE POLITICAL ENVIRONMENT

Decision making is, in a way, politics oriented administrative process. We have discussed the question concerning environment earlier wherein economic, social, and political aspects have briefly been analyzed. In a democratic system of governance, primacy of politics is often witnessed, particularly in the third world countries. In India these can be seen and the state of Uttar Pradesh particularly explains the phenomenon. In order to ascertain these view points, the respondents were asked to give their opinion about political interference in the governmental working. The choice (option of the respondents was limited to the secretariat and the field postings). The responses received in this regard suggest that about sixty percent (58.62 per cent – T. 5.3) of the respondents expressed in favour of field and about thirty per cent (29.31 per cent) in favour of secretariat posting.

The primacy of politics at different administrative levels calls for different reaction from the people who matter in the decision making. In view of this, the respondents were asked to give their impression regarding the reaction to such interference. The opinions given by the respondents show an interesting response pattern. It could be seen that majority of the respondents (77.80 per cent – T. 5.4) have opined that such interference is normal in a democratic system. A considerable percentage of respondents (17.24 per cent) chose not to reply or express their opinion on this question. Some of them expressed that they don't pay attention to such political interference.

It may be noted that during the post independence period, India opted for a neutral bureaucracy. Though of late, there has been an alleged politicization of bureaucracy. There are a number of factors which have caused this politicization. One such factor has been that the civil servants often take recourse to political approach to seek favour in the service. In view of this, respondents were asked to give their opinion as to how far they have taken recourse to political approach for favour in the service. There is about equal division of respondents who have expressed 'sometimes' and who have said 'never' or 'sometime' (44.83 per cent T. 5.5). The percentage of civil servants using political approach in the service is very small (about 7 per cent). The percentage seems considerable in either case which shows that a large number of civil servants have now taken recourse to political pressure to seek favour in the matter of their service or may be to avoid discomfiture of any sort, though considerable percentage of respondents have denied this during the course of the study.

In order to assess the political and social overtones in the administration, an attempt was made to know from the respondents about their own caste people approaching there for some personal work or favour. The respondents were, therefore asked to give their opinion as to how often your own caste people approach you for help. It could be seen from the responses received (T. 5.6) that large number of people (58.62 percent) argued in the affirmation when they said that they do 'sometimes' approach for work. During the course of study it was observed that such a phenomenon is in operation wherein caste people try to seek favour from their people in the bureaucracy. However, about one-fourth of the respondents opined that caste people 'never' approached for help. It is interesting that about 14 percent of the respondents expressed that 'very often' the caste people approach for help. In an over all analysis, therefore, it could be observed that socio-political environment of the State is such where such things have come to be accepted as normal.

While conceding the politico-socio environment in terms of favour, we wanted to assess the reaction of the respondents in such a situation. The respondents were thus, asked to give their opinion as to whether they consider such approach positive or negative. Looking at the response pattern (T. 5.7) one finds that majority of the respondents (51.73 percent) have expressed that it is negative. It may mean that they do not oppose of this approach, but accept it as a part of the prevailing social ethos in the State. As against this, about 28 percent respondents felt that it is positive. Their reaction seems to have been based taking into account the democratic framework of governance.

THE NATURE OF BUREAUCRACY

There is a general agreement on the issue that India opted for and continued with a neutral, rule-bound Weberian bureaucracy after independence. Most of the countries which were under the British Empire, after attaining freedom persisted with the similar kind of bureaucracy which was there in the earlier period. In the initial decades of modern, independent India, not much problem was perceived in the administrative working and there was some sort of cohesion between the political and administrative wings of government. But later, particularly during Mrs. Indira Gandhi's tenure, problem started surfacing in one form or the other. In this context, the prevailing nature of Indian bureaucracy was also questioned for its neutrality and anonymity. On the contrary, the idea of 'committed bureaucracy' was advocated for the speedy implementation of poverty alleviation programmes launched particularly during Mrs. Gandhi's period. However, this idea could not cut much ice and it was viewed with suspicion then by different sections of intelligentia in the country. The idea of 'committed bureaucracy' therefore, could not be pursued further.

The question of the nature of bureaucracy, however, keeps bothering in view of socio-economic and political change anticipated in the broader national interest. In view of this, we tried to ascertain from the respondents whether they would prefer a 'caste-neutral' bureaucracy. Their responses show that the majority of the respondents (55.17 percent) opined in affirmation. This would on the one hand, mean that civil servants, by and large, today are in favour of neutral bureaucracy. This may be taken as denying the fact of social upheaval, the politicization and other factors affecting the society in general and the administration in particular. As against this, we have found that slightly less than half (44.85 percent) of the respondents expressed that bureaucracy can not be caste neutral. Both these view points have their own merit and demerits. But the fact remains that now the societal changes have effected the administration in their own way.

This question was put to respondents in view of the fact that caste is playing a dominant role in the societal process.

In a subsequent question, the respondents were asked to give their opinion as to whether the bureaucracy be caste neutral. In response to this question, majority of the respondents (81 percent T.5.8.) expressed that the bureaucracy should be caste neutral. As against this only 17.24 percent opined that it should not be caste neutral. In the earlier question too, the respondents expressed that it is possible to have caste neutral bureaucracy in a caste-ridden society. The opinion expressed by the respondents might look a bit theoretical, but they did express that bureaucracy should not be influenced by the caste factor. It is, of course, another matter that in practice it can not ignore such factors and forces.

In our foregoing discussion, we have attempted to explain that in Uttar Pradesh the role of caste is predominant in the societal process. We therefore, wanted to assess as to whether the role of caste has increased or decreased over the years. There are a number of studies which have tried to examine and assess the caste in social process. We on our part wanted to know whether the role of caste over the years have decreased or it has increased in the context of the administrative process. The respondents were, therefore, asked to give their opinion on this issue.

A considerable majority of respondents (67.24 percent T. 5.9.) have expressed that the role of the caste has increased over the years. However, around 28 percent of the respondents did feel that it has decreased. This assessment seems to be in conformity with the observation we had made during the course of our field work, which has shown that the social order in the state has, by and large, become caste ridden or caste oriented. It can hardly be denied that the caste has played a role in the socio-political process in the state. We wanted to ascertain as to the nature of their role. The participants were, therefore, asked to give their opinion as to how far the caste has played a role in bringing about social change in the state. A close look at the figures (T. 5.10) would show that half of the respondents felt that caste has, to some extent played role in bringing about the social change, though a limited percentage of respondents have opined that it had largely played a role in bringing about social change. If we confine these two opinion of the responses, we find that the role of caste in bringing about social change has been accepted by the majority of the respondents, it is a question of degree where the difference is noted. However, a considerable percentage (31.03 percent) of respondents have expressed that the role of caste is bringing about social change falls under the 'not at all' category responses. This shows that still there are people who find that the role of caste in bringing about social change has not been very significant. This response category respondent may be influenced by other thought process which negates the role of caste in social formation.

WORK RELATED

In order to know about the work environment and the satisfaction level, the respondents were asked work related question. This was done to assess the situation regarding the work and the work related scenario. The answers received from the questions are discussed below.

Since the governmental decision making mainly centres round the idea of public interest, we wanted to assess from the respondents as to how far their work involves the interest of the people. In view of this, they were asked to respond to this question. The response pattern shows that about less than half (46.55 percent – T. 5.11) of the respondents expressed that up to some extent their work relates to the interest of the people. Further, a considerable percentage (15.52 percent) of the respondents opined that their work is not at all related with the interest of the people. It seems interesting that only 36.20 percent of the respondents have said that their work is largely in the interest of the people. In a recently conducted study of Uttar Pradesh Secretariat, it has been observed that officers generally remain pre-occupied in routine matters. Consequently other important functions take a back seat. In order to assist the satisfaction levels of the officer with their work and the environment, the respondents were asked to give their opinion as to how far they are satisfied with the present work. The responses received in this connection show that majority of the respondents (62.07 percent - T. 5.12) expressed that they are satisfied only 'to some extent' with their present work. As against this, the percentage of respondents who have preferred the response category 'largely' is 29.31 percent. It shows that less than one third of the total respondents reported largely satisfied with their present work. Added to this, about 9 percent of the respondents chose not to reply. A closer analysis of the figures provides us with the fact that people in the secretariat are generally not a very satisfied and motivated lot. This may be one of the reasons why the officers remain pre-occupied with routine matter rather than taking up the challenging and innovative task of governance. We have earlier referred that there is a general feeling about political interference in the administration. This interference is exercised by different political parties depending upon their own political interest and orientation. In this context, it was considered relevant to know as to whether this interference increases or decreases with change in the ruling alliance. The respondents were, therefore, asked to give their opinion whether the extent of interference varies with different ruling allies.

The analysis of data in this context has shown that there has been variation in the interference extent with change in the ruling party alliance. A clear majority of respondents (72.41 percent - T. 5.13) have expressed that the extent of interference varies with different political parties. During the course of the study, it was observed that large scale changed at different governmental levels where affected whenever there is a change of government at the political level. The respondents too have corroborated the observation made during this course of the study. However, about one fourth (25.86 percent) of the respondents did express a contrary opinion, saying that interference does not vary with change in ruling alliance.

India has progressed through a multi-party system during the post independence period. This period has seen the emergence of a number of political parties. These political parties allegedly have different value orientation and cultural bearing. This has affected the decision making process as well. In view of this, we wanted to ascertain from the respondents as to whether different political parties have different political culture. The response pattern received shows that a great majority of the respondents (84.48 percent – T. 5.14) expressed that different political parties have different political culture. As against this, only 15.52 percent of the respondents opined in the negative. In view of the prevailing political culture in the State, the administrative process has got influenced. It clearly substantiate the common understanding of the political culture prevailing in the country.

The social background plays an important role in upbringing of the individual and the formation of values and attributes. The education, economic status etc. plays a role in shaping the behaviour pattern which consequently influences the work pattern of individuals and groups. In view of this, we tried to collect information on this aspect. We asked the respondents to give their opinion as to whether the social background of the politician is reflected in their behaviour. The data on this aspect clearly substantiates the general observation that social background of the politician is reflected in their behaviour. It is important to note that majority of respondents (55.73 percent – T. 5.15) have expressed that social background of the politician reflects in their behaviour. Their social background, therefore, affirmatively affects the behaviour pattern. Also, a considerable percentage of respondents (44.82 percent) felt that only to some extent, the social background of the politicians get reflected in their behaviour. It is, therefore, a matter of fact that social background gets reflected in the behaviour of the individual or the group. A very negligible percentage of the respondents (3.45 percent) expressed that it does not at all gets reflected in the behaviour.

A similar question was asked to the respondents as to whether social background gets reflected in the behaviour as well. This was done with a view to ascertain if there was any variation in the response pattern. It is generally seen that people adopt different yardsticks in responding to social issues. The responses received show a marked variation in the response pattern to the question . In this context, slightly less than half of the respondents (48.27 percent - T. 5.16) have expressed that social background only to some extent is reflected in the behaviour pattern. Interestingly, this is the highest response as well. A considerable percentage of respondents (36.22 percent - T. 5.16) have expressed that it largely gets reflected in the behaviour.

In order to know the various factors that affect the work culture of the civil servants, respondents were asked to give their opinion on this aspect. A number of factors have been reported in this regard. Some of the important factors referred to are being analyzed here. Scholars agree that there is a relationship between the work and work culture. A cohesive work culture can produce better work. In view of this, we wanted to ascertain the satisfaction level of the respondents with the present work culture. It is worth noting that the highest responses to the question is reported in 'to some extents' category (68.97 percent - T. 5.17) only a negligible percentage (8.63 percent) felt 'largely' satisfied. As against this a considerable percentage (20.68 percent) have reported that they are 'not at all' satisfied with the present work culture.

In view if the prevailing work culture and its impact on the work, we wanted to ascertain as to whether there is a need to change the prevailing work culture. The response pattern clearly indicates the positive desire for changes in the prevailing work culture. A large majority of respondents

(60.34 per cent – T. 5.18) have expressed that there is a need to change the present work culture. The next highest response category (34.49 per cent) reported 'to some extent' which also have favoured the need to change. Only a negligible percentage (5.17 per cent) of respondents have opined that there is no need for change in the present day work-culture.

As has been mentioned earlier, the administration has often pre-occupied itself in the posting and transfer of the civil servants. It is thus, important to note the various factors that affect their work conditions. In order to ascertain the intensity of various factors, the respondents were asked to give their opinion on these aspects. The responses show an interesting pattern. Various considerations guide the administration in postings and transfers. In the above response pattern, political consideration account for almost one-fourth (25.86% - T. 5.19) of the total respondents. Caste considerations too play a role as 10.34 percent respondents opted for this category. The highest percentage (31.04 percent) though not very high has expressed in favour of merit as an option. It is quite interesting to find that 17.24 percent of the respondents opined money as a consideration in the postings. In earlier days these considerations were unheard of whereas in the contemporary administrative scenario, considerable percentage of respondents have opined in favour of this factor. Over all, therefore, we fined that external factors have influenced the administrative process in which we have tried to identify some of the factors in our administrative activity which is performed at the Secretariat level.

ATTITUDINAL QUESTIONS

Attitude of an individual generally conditions his response pattern and these attitudes are formed over a period of time. The individuals interaction within the family and the society at large help in attitude formation which goes to determine his value frame and attitude towards things. Since administrative process is part of the broader societal process, it is considered important to note the attitude of the people in the administration. In view of this, we have placed some relevant question to determine the attitude of the respondents towards socio-political and administrative issues. We have also collected some biographic information about the respondents which is being presented here.

Of the respondents selected for detailed information, 58.62 per cent were having rural background whereas 39.65 percent belonged to urban areas. So far as the schooling of the respondents is concerned, the information collected in this regard shows that the respondents have been fairly distributed so far as their schooling is concerned. A considerable percentage of respondents (39.66 percent, T.-5.20) had their schooling at the government run schools while an equal percentage (17.29 percent) is reflected for the village schools and public schools respectively. Almost one tenth of the respondents are reported to have their schooling at private schools. Information about the schooling of their children was also collected. The data shows that the highest percentage (39.66 percent, T. 5.21) is reported in the category of private school and the next higher percentage (27.59 percent) is reported in the category of public school. The government schools account for only 25.86 percent. By comparing the figures about schooling of the respondents and their wards, we find the different response patterns. It is interesting to note that considerable percentage of respondents did have their schooling in the villages and municipal schools. But their children are attending public and private schools. These schools are offering education at a much higher cost than those being offered at the municipal or village schools. This marks an attitudinal shift in schooling.

The issue of reservation is being debated from various angles. Many people have been arguing that the reservation has affected the merit in the civil service while others have been arguing in favour of reservation. We, therefore, wanted to ascertain from the respondents their opinion on this issue. The information received shows that a very small percentage (13.79 per cent T. 5.22) of respondents expressed that reservation helped in restoring merit. It is worth noting that fifty percent of the respondents opined that to 'some extent' it helped in restoring merit. Against this background, we find that considerable percentage (34.48 percent) of the respondents expressed that reservation did not help in restoring merit in the civil services. It can, however, be noted that people who have taken advantage of the reservation policy may like to opine that it has restored merit while others may like to express a contrary opinion.

CONSIDERATIONS THAT GUIDE YOUR ADMINISTRATIVE DECISION

We have noted earlier that various factors influence decision making in the government. In view of this we have tried to address this question to the respondents to know their opinion on this issue. There are several considerations which guide the civil servants in taking administrative decision. The responses received clearly indicate various factors and their intensity that influence administrative decision. About equal percentage (10.34 percent - T. 5.23) is reported for political and social consideration in administrative decisions. It is quite significant that a high percentage (68.97 percent) respondents have reported that rules on the subject are the major consideration in administrative decision. This has been the consideration for many political systems which have democracy and value neutral administration. But, we have also noted that in the present administrative scenario external factors have influenced the decision making and the governing process though the degree of influence varies.

We have already noted that caste considerations do play a role in the societal process. Opinion however, varies when it comes to identify the role of caste in administrative process. We therefore, addressed this question to the respondents as to how far the caste factor influences the administrative process. When the question regarding caste consideration was addressed to respondent, not many have reacted favourably to this question. Now, they have opined in favour of caste-dominance in the administrative process. A considerable percentage (32.76 percent - T. 5.24) of respondents have expressed that it largely dominates the administrative process. Also, majority of the respondents (63.79 percent) felt that to some extent only caste dominates the administrative process. Very few respondents (3.45 percent) felt that the caste not at all dominates the administrative process. Taking an overall view, one can argue that caste plays a dominant role in the administrative process of the state.

Since the caste has a role in the administrative process, it is important to know as to how far this helps in solving the problem. In view of this, the respondents were asked to give their opinion as to how far caste factor help in solving the problem. The response pattern to this question shows that majority of the respondent (62.07 percent - T. 5.26) have expressed their view that the caste factor does not help in solving the problem. However, a very negligible percentage (6.89 percent) of respondents felt that it largely helps in solving this problem as against 13.04 percent respondents who opined that it only to some extent help in solving the problem. The predominant view therefore, has been that the caste factor does not help the administration in solving the problem.

In the initial days, the bureaucracy presented a homogeneous group. However, over a period of time, changes have taken place in the composition of bureaucracy. It is now generally seen as a heterogeneous group. It is to be seen as to whether the heterogeneity of the group also gets reflected in the attitude. In order, therefore, to seek explanation to these queries the respondents were asked to give their opinion as to whether the heterogeneity of the bureaucracy gets reflected in the attitudes. The response to this question clearly shows that over the years some changes are getting reported in the bureaucrats' attitude. About one fifth of the respondents (18.96 percent - T. 5.26) have opined that heterogeneity in bureaucracy is largely getting reflected in the attitudes of the civil servant. As against this, very few respondents (8.63 percent) felt that heterogeneity does not get reflected in the attitudes. Majority of the respondents (72.41 percent) have expressed that heterogeneity to some extent gets reflected in the change of attitudes of civil servants. It can, therefore, conclucively be observed that hetrogenic nature of the civil service in the state gets reflected in the attitude of civil servant which has shown some change. As can now be seen that heterogeneity is affecting attitudinal changes in bureaucracy, it can be interesting to see its influence in the social process. It is agreed that different people in bureaucracy have different attitude towards social questions. This may help in bringing in social cohesion. In order to assess these value premises, the respondents were asked to give their opinion as to how far different attitudes help in promoting social cohesion. The response pattern to this question shows that majority of the respondents (62.08 per cent – T. 5.27) have expressed that to some extent it helps in promoting social cohesion. About one fifth of the respondents, however, fall in the category of 'not at all' or 'no response' while some of them have chosen not to reply as against others who felt not at all. Interestingly, about one fifth of the respondents have expressed that it largely helps in promoting social cohesion. In view of the response pattern shown here, it is difficult to make a clear observation regarding the role of divergent attitudes of the civil servants in promoting social cohesion.

The general impression about bureaucracy is that it is averse to change. However, heterogeneity and other factors have allegedly brought about some change in earlier attitudes. In order, therefore, to assess whether the attitude of bureaucracy is accelerating growth and change, the respondents were asked to give their opinion on this critical question. The responses received indicate interesting trend about which some people have their own reservation. It is important to note that one third of the respondents have expressed that the attitude of bureaucracy has largely helped in accelerating growth and change. As against a considerable percentages (15.52 percent - T. 5.28) of the respondents felt otherwise. In their opinion, the attitude of the bureaucracy 'not at all' helped in accelerating growth and change. However, about half of them have expressed that only to some extent the bureaucratic attitude helps in accelerating growth and change.

We have earlier made an observation that the nature of the society is heterogeneous. The heterogeneous character of the society gets reflected in the administration. We, therefore, wanted to know as to how far the heterogeneous nature of society affects the administrative decision making. In this light, the respondents were asked to give their impression. The response pattern indicates a positive influence of the heterogeneous nature of the society. A considerable percentage of respondents (27.58 percent - T. 5.29) have expressed that they largely affect the administrative decision making. Majority of the respondents (68.97 percent) opined that the influence of the heterogenetic nature of society affects the decision making only to some extent. Only a negligible percentage of respondents (3.48 percent) have expressed a contradictory opinion. In an overall analysis, therefore, we find that there is a positive tone so far as the influence is concerned, the variation is only in the degree of that influence.

A general perception has been that administration has an elitist nature which has adversely affected the developmental process. We, therefore, wanted to reassess the situation as to whether any change in this perception has been affected by changing societal factors. The respondents were therefore, asked to give their opinion on this question. The information received in this regard shows a response pattern which explain that earlier perception continues to hold good. A considerable percentage of respondents (27.58 percent – T. 5.30) have expressed that the alleged nature of bureaucracy has affected the development process. Also, majority of respondents (68.97 percent) opined that to some extent this has affected the development process. It is only a negligible percentage (3.45 percent) which felt otherwise. On the whole, therefore, we get an indication that elitist nature of bureaucracy is still a factor affecting development.

In continuation to the earlier debate about heterogeneity, we wanted to assess from the respondents as to whether this gets reflected in the bureaucracy as well. It is interesting to note that the responses received in this regard show an affirmative trend. The heterogeneous nature of the society is now reflected in the bureaucracy. A considerable percentage of respondents (22.41 percent - T. 5.31) have expressed that the heterogeneous nature of society is getting reflected in the bureaucracy as well. Also, a majority of respondents (68.97 percent) have felt that 'to some extent' it is reflected in the bureaucracy. Very few respondents (6.89 percent) have opined in favour of 'not at all'. Taking an overall view, one find that this is a positive linkage between the society and the bureaucracy which the respondents have expressed.

It is generally felt that some change has been observed in the texture of bureaucracy. The heterogenetic nature of society is allegedly being reflected in the bureaucracy as well. We, therefore, wanted to ascertain from the respondents as to whether it is helping the cause of development and growth. It could be seen from the response pattern that about one fifth of the respondents (T. 5.32) expressed that it has not largely helped the cause of development and growth. However, slightly more than half of the respondents opined that only to some extent it helps the cause of development. The interesting part of the response pattern is that about one fourth of the respondents have somehow felt that it has not at all helped the cause of development and growth. This opinion, therefore, has to be taken into account while making an assessment as to whether heterogeneity helped the cause of development and growth in the society.

We have so far presented the analyses based on the data collected through the structural questionnaire. We now propose to take up a brief analysis of the based on our observation and discussion had with selected officers during the course of our field investigation. In the process of this analysis, we may again refer to the earlier analysis by way of comparison and supplementation. Also, in this analysis we would take up only selected areas / issues which are directly concerned with the objectives of the study to avoid repetition.

During the course of our field survey three major areas referred to were concerned with the services, the work and the attitudes of the officers. In the light of these, we guided our observations of the administration. It has generally been observed that there has been an increasing interference in the day-to-day affairs of the state. The officers confessed in private that due to excessive politisation of the administrative system, the officers and governance is suffering. It is however, important to note that in a democratic polity, administration has to take political leadership into confidence. But, over the years, the political leadership is trying to indulge in routine administrative matters which many a times results in 'role crisis'. This role crisis is observed in the decision making process and governance. Since the politicians are continuously bothered by their partymen and other similar groups for getting their work done, these politicians in turn direct the civil servants at different levels to help them out. These are generally matters regarding seeking some favour or the other for the coterie. Often, caste people are placed on important positions down the line for various kinds of patronage. The caste factor has also been observed during even routine postings and transfers. These extraneous factors have influenced the administrative process adversely. Efficiency and effectiveness of the administrative machinery have suffered and general apathy is visible in the administrative working.

It has often been observed that honest and upright officers get neglected and isolated in the present rut of administration. Their motivational level too seems to be rather low. They remain generally reluctant to offer frank and fearless advice as the same may not be liked and appreciated by the political masters. They even may be punished for such kind of advises. In view of these, the civil servants generally opt for the path of least resistance, although that may not be in the best interest of the people of the state. This is however, a situation which has often been observed during the course of the field work.

Different political parties adhere to different political culture. The major political parties in the state for some time now are the BJP the Samajwadi party, the B.S.P., the Congress apart from a few others like Janta Dal, the Communists who have little influence in the political and administrative process. The BJP has a different political culture which is rooted in their ideological commitment. The Congress, which is one of the oldest political parties, has its own ideology. Although, over the years, some changes are seen in the commitment of the Congress party, but by and large, it has continued with its own culture which is different from that of the BJP. Some newer political outfits like Samajwadi Party and the Bahujan Samaj Party have their cultural mooring which again is quite different from that of the Congress or the BJP. The varying political cultures had their consequential impact on the behaviour pattern of the political parties and their bosses. Depending on these, the extent of political interference in administrative matters has been observed.

The Secretariat of the State is the apex decision making body. It is important to note that various government policies, programmes and decisions involve several ministers. The State Secretariat also functions as the channel of communication with the Central Government including Central bodies like Planning Commission, Finance Commission, apex financial institutions, Election Commission etc. This makes it necessary for the secretariat to have an intimate knowledge of national policies, programmes and prospects. How far they have that knowledge is a matter of opinion as sometimes we have observed lack of these during the course of our field survey.

The government is run on the principle of collective responsibility of cabinet and while the views of the individual minister may differ in a particular matter, a collective decision has to be taken in the interest of the State. Thus, a 'collective consciousness' has to be generated. This is probably the most difficult task to be handled at the Secretariat level and the adequacy or capability of the Secretariat organization, is ultimately judged by the speed with which interdepartmental conflicts are resolved or sorted out and decisions taken. However, during the course of our field work various situations have been observed. An interesting example in this regard is that State Government is unwilling to accept the recommendations of the Ninth Pay Commission for increasing the retirement age from 58 years to 60 years as this would allegedly affect the interest of the OBC in the State. Whereas, the government implemented the recommendations regarding the pay structure of the Pay Commission, it is finding difficult to implement the recommendations regarding the age group. In a democratic system, such pressures are bound to occure and influence the decision making in the Government. The interesting part, however, is that the government is less concerned about the general interest of the people and more about its constituents, the parties and the groups.

During the last few years, the State is being run by a coalition of parties, sometimes support from out side the House and sometimes being part of the Government. This coalition run government has been through the period our study as well, and therefore, we have tried to observe the various pulls and pressures of these governments on the decision making process. It has been observed that coalition is a very ticklish game of power. Various constituents of the coalition heading various ministries, try to concentrate power in their own hands while the Chief Minister's Office would prefer to have a centralized control and governance as their own power, status and prestige remains at stake. It, therefore, turns out to be a continuous trigger for power for various constituent in the coalition government. This is more acute and apparent in important ministries.

In view of the above political scenario, it is difficult to

pinpoint the priorities of the political matters as well as the civil servants so far as the decision making process is concerned. One important fact which keeps haunting there thought process, the uncertain and short tenure of their government. They, therefore, become more concerned about the short-term political and personal gain than the long-term gain for the State and society at large. One can quote several examples in support of this observation which had predominantly been seen during the course of our field survey. A number of smaller political outfits which become part of the coalition government concentrate more on consolidating their political background. One the one hand, it could be considered as natural, on the other, it has adversely affected the development of the State owing to an unsettled political scenario. Its impact has largely been visible on the administrative system as well.

The question of attitude has been another important concern in public administration where decision making is the major activity. The attitude of political system towards the civil servants and that of civil servants towards the general mass of the people remains a core concern in the decision making process. In the earlier part of the book, we have presented an analysis of this attitude and its impact on decision making. Different attitudes have been observed in the politicians as well as civil servants towards each other which at times has been not very congenial and cohesive. The political masters exhibit a very demanding attitude owing to shorter span available to them which causes anxiety among the civil servants. Further, it is not only question of politicians and civil servants towards their colleagues. Since there is hierarchy amongst the civil servants, there is variation in the attitudes of the civil servants towards their senior and junior colleagues. The factors which have conditioned their attitude include the caste factor apart from other social factors. This has caused almost a divide amongst the civil servants which may not be visibly seen but intrinsically this has been observed. The way officers belonging to different caste groups encourage or discourage people evidently shows their preference in inter-personal relationship. It can therefore, be submitted that different attitudes have been witnessed amongst the civil servants which has consequently affected the decision making process as well.

In the present chapter, therefore, an attempt has been made to make an analysis of the role and influence of caste factor in the decision making process with particular reference to U.P. Secretariat. The analysis bases itself on primary data collected through a structured questionnaire alongwith observation spread over the field survey of the area.

Chapter 6
The Impact Analysis

The post Independence India opted for a system of governance which based itself on the principles of democracy and secularism among other related ideas. There has been an in-built primacy of the rule of law in the decision-making and the governing process. We have made an attempt in the earlier part to show that there are several factors influencing the decision making and the governance including the caste factor. We now propose to briefly analyze the impact of these external factors on the decision making as well as the overall governing process with particular reference to Uttar Pradesh.

It is now being realized that there is a causal relationship between administrative efficiency and the value-premises of decision making process. The value premises and the quality of decisions influence the governance and administrative efficiency. We have witnessed these factors and their impact on administrative efficiency, accountability and the quality of governance. There is a telling impact of the external factors including caste in the overall decision making and the administration. Uttar Pradesh is a multi-religious, multi-cultural society. It is also the most populous and a background state of the Indian Union. The social order is such where caste plays a predominant role which now extends to the administrative arena. The extended role of caste is socio-political and administrative process does not seem to have done any good in the growth and development status of the states which continues to be a matter of concern.

In case we agree that there is a continuance in the socio-political and administrative processes, we have to look into the priorities of the political parties, and, political bodies, who make the ultimate decisions which shapes the polity and the governance. With this in view, if we have look into the political parties and the political leaders who have been at the helm of affairs in the state the point becomes rather clear. For the last couple of decades the political leadership at the helm of the affairs at the state level has been more keen to stabilize their political (personal) base rather than devoting time, energy and concern for matters of general public interest and welfare. At times an impression is generated that government of the day is having only a sectarian interest rather than general people's interest. During the last fifteen-twenty years the state has seen the Chief Ministers in the name of Kalyan Singh, Ram Prakash Gupta, Rajnath Singh, Ms. Mayawati, Mulayam Singh and they headed a coalition of Parties. Some of them have been Chief Ministers for more than once. Interestingly each time a particular Chief Minister takes over the reign of office, there is a marked change in the air, in the bureaucracy and otherwise also.

The caste factor in Uttar Pradesh is a predominant factor in the governance. During the course of our study we came across the views of the bureaucracy on this issue which they generally accepted. Recently, the city of Lucknow witnessed a clash between the advocates and the police which marked a nationwide protest by the lawyers. The caste factor here too is being witnessed. For example, it has been reported: 'Don't send a Brahmin reporter to cover the developments in the strike, requested a senior member of the Bar referring to High Court lawyers agitation in Lucknow following a lathi charge recently'[1]. Nothing could be more telling than this request about the prevailing attitude in the **Bimaru** state of Uttar Pradesh. What is important here is that it is not a politician seeking votes in the name of caste but a lawyer supposed to be an intellectual, bearing this narrow mindset.

It is not just the voters in the villages but also the city-bred, educated people who see nothing wrong in thinking on caste lines, so what if it perpetuates an obnoxious practice that only help to divide the society. The lawyers in Allahabad High Court are just as deeply divided in Brahmin, Thakur and Kayastha lobbies as the polity and they can go to any extent to assert their castiest, and professional superiority[2]. Once can find such potent divisions in others vocations as well like education.

It was natural therefore, that when a member of the Department of International Development (DFID) wanted to know what are the emotive issues pertaining to good governance that would excite the people in Uttar Pradesh one fumbled for a reply. Caste, which has nothing to do with development, it appeared, was the biggest emotive issue that could bring people together. Illiteracy, health care, potable water, power supply, infant mortality, poor infrastructure, corruption are issues that do not move us into a frenzy like mandal or a temple movement could do. The question one would generally ask is why people don't react to lack of basic amenities, to all pervasive corruption, for employment and against casteism and communalism? Perhaps, they are too docile to give went to their feelings. Or, perhaps they have accepted bad governance as fait accompli.

During the last decade or so Uttar Pradesh has witnessed different Chief Ministers who demonstrated different styles of functioning. In a recent news item in a local daily Uttar Pradesh has witnessed the caption published was "Governance Mulayam Style'[3]. It elaborates that if the allocation of work among bureaucrats is any indication, there is utter chaos in Mulayam Singh Yadav government. While some officials are holding posts more than one, there are others who have been either kept in waiting or given post with hardly any work. The present Agricultural Production Commissioner (APC) and Additional Chief Secretary Neera Yadav, holds two portfolios. She is the APC and also the Principal Secretary, Education. The APC has thirteen departments which include mega sectors like agriculture, rural development, panchayat raj and dairy. To hold the responsibility of all these important departments is an extremely strenuous full time job for any senior bureaucrat. But Neera Yadav is also holding the charge of Principal Secretary, education encompassing primary education, secondary education, higher education and technical education. This sector too carries immense responsibility and tremendous workload. In the past there have been instances where the different departments of education alone had two Principal Secretaries. Senior officials admit that it is humanly impossible to do justice to both the portfolios even if one works twenty four hours a week. The political analysts have observed that no chief minister in the recent past has looked at the people of the state as a whole. They are looked as either constituency or castes, or as Hindus and Muslims[4]. Only a few months ago, Raghuraj Pratap Singh alias Raja Bhaiya became the rallying point for all the Thakurs of U.P., led by Samajwadi Party, Thakur Amar Singh. BJP's Rajnath Singh tried his best to win over the community but his party's long association with the BSP dented all his efforts because a Dalit Chief Minister had hurt the Thakur pride. Like Thakurs, Brahmins, Yadavs and Dalits have had their highs and lows depending upon the castes to which Chief Minister of the day belogned[5]. Recently in an apparent infringement of the set norms, the state government nominated ex. State Minister and legislator from Gasari (Balrampur) Shiv Pratap Yadav as chairman of the Uttar Pradesh Finance Corporation (UPFC) Kanpur. The issue has snowballed into a major controversy with the Small Industries Development Bank of India (SIDBI) authorities questioning the state governments', supposedly unilateral decision to appoint a political figure against the precedent[6]. A lengthy correspondence is reported between the SIDBI and the government on the issue.

The dominant castes in the state had their own share in the governance which has kept swinging. If we look back to the year 1967, we find a drastic change, in representative politics of Uttar Pradesh which had its consequential impact on the decision making. That year, a short, stocky young man made his debut in the State Assembly. He was at that time barely 27, till then the youngest MLA ever to be elected. An agitational politics in his blood, this young man was distinct to play a much bigger role than perhaps he himself may have visualized. We are talking of Mulayam Singh Yadav[7]. Since then Mulayam Singh Yadav have been able to sustain and increase his influence in Indian politics as well as his control in the governance of Uttar Pradesh. It is ironical that so called national political party have gradually margianalized and the regional parties like S P and B S P have gained sufficient ground in the socio-political process of the state.

When Mulayam Singh Yadav entered the political arena in 1967 he was a child of the first real consolidation of the anti-Congress vote. The election marked a para-digimal shift in terms of existing political behaviour through out the North and of course, in Uttar Pradesh as well. As Gillian Wright (Seminar, August 1995) perceptively points out "... essential to Congress domination was the division of its opponents. Due to lack of education, organization and economic power, no separate Dalit movement could develop. Neither could Jan Sangh expand its vote much beyond urban and trading communities (many traders preferred to vote for Congress) and the traditional upper caste elite of Awadh. Land reforms robbed the community of the issue most likely to give them a base. There was no united backward vote either. Upwardly mobile members of the middle peasantry, especially Jats and Yadavas were naturally inclined against the Congress. This was because, after Charan Singh spearheaded the Jamindari abolition legislation in 1951, they and the former landed elite were rivals for political power which gave control over economic resources. As that elite was in the main attached to Congress, the middle peasantry looked elsewhere".

In terms of our central concerns, we have tried to focus on the phenomenal rise of bullock capitalist- a very expressive term which signified both social and economic aspects of the brash post-jamindari, upward mobile, peasant proprieters[8]. It is this category which has symbolized the changed political contours in Uttar Pradesh. Force and coercion compelled

reservation have helped the target groups or the society in general. While it is true that benefits of reservation came to be appropriated by the upper segments within the backward caste (for greater elaboration, see A.M. Shah's excellent essay: 'The judicial and sociological view of Other Backward Caste, in caste: in Twentieth Century Avatar, Viking, 1996). There is no doubt that reservation, in its initial stages, helped SCs to compete better[15]. Interestingly, most dalit leaders are not very convinced about this fact though Dr. Bhimrao Ambedkar wanted the reservation policy to be just a temporary measure, and Kanshiram has on many occasions lashed out at this policy.

The value premises of modern governance has generally focused on value-neutral paradigm. The discussion in the forgoing pages would suggest that the decision making process and the governance in Uttar Pradesh for quite sometime has been partly, if not generally, caste-oriented. Taking cognizance of caste and basing governmental decisions on caste could, to an extent, hamper the interest of other social groups. This is also, to an extent, denying the various social formation, other than the ruling elite, a share in the governance and empowerment. The impact of such governance ceases to be good governance, it operates on limited, factional interest, affecting the broader societal interest. The state of Uttar Pradesh has been suffering from such short sightedness of political leadership who refuse to see beyond the caste or the class formation. Such governments failed to appreciate the problem faced by the general mass of people and therefore, general governance suffers.

It is worth noting that successive governments have made tall claims of making the governance in the state people-centred. But each term they have failed to keep their electoral promises and one prime reason could be the narrow political or parochial interest. The most recent addition to his to the governance is that of the Samajwadi Party headed by Mulayam Singh Yadav who promised to make Uttar Pradesh— "Uttam Pradesh", a great state. However, the ground realities do not support the claim made by the Chief Minister. We have already made our observation in the earlier pages about the caste-oriented governance in which the governance itself has become a casuality: the impact of the caste in the socio-political process has yielded a fractured social order which often failed to appreciate the 'totality' of the governance for which it is elected and is accountable.

REFERENCES

1. Atul Chandra, 'Caste as an emotive issue', *Times of India*, Lucknow, 19.4.04.
2. *Ibid.*
3. *The Times of India*, Lucknow, July 7, 2004.
4. Atul Chandra, *op. cit.*
5. *Ibid.*
6. Akhilesh Kumar Singh, *Times of India*, August 1, 2004.
7. Lal & Nair, Caste vs. Caste; Turbulence in Indian Politics, Ajanta Books International, 1998, p. 32.
8. *Ibid.*, p. 11.
9. *Ibid.*
10. Debashish Mukarji, Kaulaiah Bhelari, Deepak Tewari and Ajay Upreti, "High Tied of Power", *The Week*, October 3, 2004, p. 16.
11. *Ibid.*
12. *Ibid*. p. 17.
13. *Ibid*. p.
14. *Ibid*. p. 18.
15. *Ibid*. p. 19.
16. *Ibid*. p. 42.

Chapter 7

Concluding Observations

The entire governing process bases itself on the decision making. We have therefore, tried to present a theoretical discussion followed by its operative angle. The factors of value neutrality have been analysed and other factors including caste has been assessed in the decision making with a particular reference.

Decision making is an activity on which depends the efficiency and effectiveness of organization. In view of its importance, an independent theory of decision making has been evolved in public administration by Simon and others. In evolving the theory of decision making, the concept of 'fact and value' premise forms the core concept. Further, in the evaluation of decision making process, the concept of rationality too occupies a prime place. In the field of public administration, the decision making has to be such which facilitates and encourages efficiency and effectiveness. In the development of the concept therefore, 'rationality' and value neutrality have been of prime concern. The initial focus on these concepts rested on the consideration that public administration should be neither arbitrary nor discriminatory. In the later development, these premises were questioned on various grounds. In this process, decision making also got influenced. Several extraneous factors started influencing the decision making process. In view of these developments it was thought imperative to examine these new value premises.

Amongst the various extraneous factors that were influencing the decision making in government, caste has been considered as one of the core factors. The role of caste in the societal process had already been studied rather extensively. Studies have also come up describing its role in the political process including the electoral. We on our part wanted to examine its role in the administrative process as it was trying to desettle the established norms in administrative behaviour. Since administration is operating in an environment which is social, political and economic, the administrative processes are bound to be influenced by such factors. In view of this, we undertook this study on caste factor in decision making amongst the civil servants of Uttar Pradesh.

The major objectives which have been explored and analysed in the book include the following amongst others:

- To identify the role of caste in administrative decision making.
- To study the linkage between caste and administrative performance.
- To study the pattern of postings and transfers.
- To study the social background of the administrators.
- To assess whether feudal values and orientation affect the decision making process.
- Whether caste overtones affected growth process adversely.

SCOPE AND METHODOLOGY

The field work for the study on which the book is generally based was limited to U.P. Secretariat and other field offices in Lucknow. The state of Uttar Pradesh is termed as mini India as it is inhabited by people belonging to different religion and castes. This social composition is also reflected in the administration. The representation of various caste groups though has kept changing over the years.

Various levels of administration are involved in the decision making process. In view of this, various factors have to be kept in mind while studying the organization, the officials and other respondents. The sample for the field study was drawn from the officers of the Indian Administrative Service, Provincial civil Service and other Subordinate Services posted in the Secretariat or other offices at Lucknow.

For the purpose of data collection we made use of both the primary as well as secondary sources of data. We have made use of observation and discussion as a major tool of data collection to help formulate opinion and arrive at conclusion. For eliciting information on different aspects of the issue a structured questionnaire was designed and administered on the officers drawn from the sample of IAS, PCS and other officers of subordinate services posted at Lucknow. The information so gathered and data so collected has been processed through the help of computer and derivative tables were prepared which are given at the annexures.

The present book basically deals with the subject of decision making amongst the civil servants and the governance in Uttar Pradesh. In view of this, a brief sketch of the state and the respondents has been presented in the book.

Uttar Pradesh is one of the largest in terms of population as well as the border state of Indian Union. Area wise it is the fourth state in the country after M.P., Rajasthan and Maharashtra. According to 2001 census, the total population of the state is 16,60,52,859 out of which 874,66,301 are males and 785,86,558 are females. About 80 per cent of the population is still rural. The comparison of decadal growth of population in the state shows that it is higher by 1.97 percent. It is interesting that every sixth Indian belongs to Uttar Pradesh.

As per 2001 census, the literacy rate in the state is reported to be 61.03 per cent. The percentage of literacy rate among male is 76.46 per cent which is more than double the percentage of literate women i.e. 44.34. The percentage of literate in rural areas is lower (36.66 per cent) against literacy percentage in urban areas (61.00 per cent). Viewed sex wise in rural and urban areas, the rate of literacy among males (52.05 per cent) and in the rural areas is more than double the rate of literacy among females (19.02 per cent). But there is not much of difference between the male literacy (69.98 per cent) and female literacy (50.38 per cent) in urban areas.

THE SECRETARIAT

The sample of the study was generally drawn from the U.P. Secretariat and other offices located in Lucknow. The origin of the U.P. Secretariat goes back to the British days when a Lieutenant Governor was appointed in 1801 at Bareilly and was allowed a Secretary and some clerks. Subsequently another Secretary was appointed in 1834 on the post. Much of the local records were destroyed in the fire which broke out in 1882 at Allahabad and there is very little material available on the early organization of the Secretariat. The earliest proceedings which are available and which are originally maintained at Fort William at Calcutta are for the year 1836 and 1837. They relate to the Foreign and Home Department of the East India Company.

The first reorganisation of the state secretariat was undertaken in 1883 and some of the important changes made in the following years were that two additional Secretary designated as Financial Secretary and Judicial Secretary were added to relieve the Chief Secretary of the excessive work. Reforms were again undertaken by the Government of India Act 1919 which were introduced in 1921. A revised 'Secretariat Manual' which contained comprehensive instructions on office organization, disciplines and office procedure was brought out in 1925.

With the establishment of Legislative Council at Lucknow in 1921, the Governor also made his residence here. The department of the Secretariat had also started shifting to Lucknow and the exodus was completed by 1935.

After independence, the State took on itself the paramount task of the development of its economic resources. The normal growth of the Secretariat and the field department took place on this account. The initial decades after independence could not witness much change in the organizational structure of the Secretariat. of the State. However, in its journey over time, especially during the last 2-3 decades, there has been a sea change in the composition and functioning of the Secretariat. The number of departments initially 48 has now increased to 74. The increase in the size of the Secretariat has created a number of problems as well.

THE RESPONDENTS

For the purpose of collecting information from the field we selected officers belonging to IAS, PCS and other Subordinate Services. The ratio of the respondents selected for interview has been 29,26 and 45 percent respectively. Of the selected respondents about 91 percent were males and the rest 9 percent were females. The agewise distribution of the respondents has been that about 5 percent were below 35 years, about 42 percent belonged to the age group of 36 to 44 years, about 36 percent were in between 45 and 52 years and the rest 17 percent were above 52 years of age. So far as the education level of the respondents was concerned, the lowest percentage of 5 were graduates, 52 per cent were post-graduate and the rest 43 per cent had professional qualification.

Information about the length of present posting was also collected from the respondents. It is interesting to note that about 35 per cent of the respondents had less than a year as their tenure. As against this, we find about 36 per cent of the respondents had more than three year as their length of service which should generally be the case of the subordinate service officers. About 12 per cent reported between one and less than two year as against 17 per cent which said more than two and less than three years.

DECISION MAKING: PROCESS – VALUE FRAME

Governance is a process which bases itself on the decision

making. In democracies governance has to be people-central which makes the decision-making an important element. The quality of decision making would determine governance. Decision making is an integral part of the management of any organization. In view of its importance in organizational operation attempt has been made by scholars to study its various facets. Decision making is generally seen as an activity or action based on calculation, reasoning and purpose. The neglect for decision making deliberate or otherwise, can cause damage to the system.

Driven home by this empirical observation, Simon maintained that decision making endeavors as the main function of an administrative organization transcending all levels and modes of the administration. Simon's analogy of decision making covered largely the psychological, sociological and political problem of every form of administrative organization. In discussing the subject of decision making it is necessary of focus on one or more of the three things: (I) The decision making process, (ii) The decision maker, or (iii) The decision itself. Simon, the main contributor in decision making thought says that decision making comprises three principal phases: finding occasion for making decision, finding possible course of action and choosing among course of action.

SCOPE OF DECISION MAKING

Decision making can occur at several levels. The first and perhaps the most basic level is that of the individual acting to satisfy the basic needs. In complex social situation individuals usually find need satisfaction as member of groups that have particular purpose. Therefore, group decision represents more than just a collection of the decision of the individual members. The scope of the decision making does not stop at the level of the group. The scope of the decision making further extends beyond the managerial process at the level of the organization. The scope of decision making is indeed wide.

RATIONALITY IN DECISION MAKING

Decision making process is also surrounded by a debate about the rational and non-rational decision making. On the one hand, the term is often applied to the choice itself – i.e. the decision is rational or non-rational. On the other hand, it is frequently used to describe only the behaviour of the decision maker.

The concept of rationality is subject to widely varying interpretation. According to Simon, "Rationality is concerned with the selection of preferred behaviour affirmatives in terms of some system values whereby the consequences of behaviour in the formal organization is objectively, subjectively, deliberately or overall rational according to the values or goals of the individual.

Rational behaviour in the decision making process simply involves the evaluation and selection of some relevant attention based upon a perceived advantage accruing to the decision maker. All that is necessary to make the choice a rational one the existence of an objective and the selection of some alternative that promises to meet the objective as seen by the decision maker.

FACT AND VALUE IN DECISION MAKING

Fact and value paradigm in decision making has particular significance. A conscious debate had been there on whether the decision be based on fact or value. Simon illustrates the harmonious blending of value and fact as the structural requirements of decision making. He says that "in so far as decision leads towards the selection of final goals, they will be called value judgements. So far as they involve the implementation of such goals, they will be called factual judgements". In administration, the mixed character of the ethical 'givens' is usually fairly obvious.

The fact and value premise are related to means and ends respectively. In the process of decision, those alternatives are chosen which are considered to be appropriate means for reaching desired ends. Thus, rationality has to do with the construct of means – ends chain of this kind. However, the means-end distinction does not correspond to the distinction between fact and value.

THE ENVIRONMENT OF DECISION MAKING

Organizations do not exist and operate in vacuum. They are a part of larger aggregate, such as economic system, social system and the political system of the nation / state. Therefore, the decision-maker must take account of the inputs of the environment in arriving at a decision and they must consider the effect or the choice on the manifold forces comprising the environment. The effectiveness of the implemented decision is conditioned considerably by the environment.

In the earlier part a detailed explanation about the interaction of the decision making process with that of the economic system, social system and the political system has been presented which shows the importance of the role of environment in decision making.

CASTE FACTOR: ROLE AND INFLUENCE

Several factors influence the decision making process. Some of these factors have already been explained. Let us now take up the caste factor, its role and influence in the decision making process. This explanation is based on the field survey conducted. We used a structured questionnaire to elicit information on various aspects to enrich the base of our data.

The information was sought from the respondents as to which posting i.e. the secretariat or field, they liked most. It is interesting to note that majority of the respondents expressed that they liked the field posting most. They have also expressed that field posting is difficult than that of Secretariat when asked about this.

Decision-making is in a way politics oriented administrative process. In a democratic polity primacy of politics is often witnessed, particularly in the third world countries. In India too, these can be seen as the sate of Uttar Pradesh particularly explains that phenomenon. To assess the extent of political interference in administration, the opinion of the respondents was sought. It has been alleged that the extent of political interference is more at the field level than at the Secretariat. Most of the respondents have considered that political interference is normal in a democratic form of governance.

In order to assess the political and social overtones in the administration, attempt was made to assess as to how far caste people approach for help. The response pattern in their opinion shows that hardly one fourth of the respondents reported never, otherwise the rest of the respondents expressed either very often or sometimes. It is interesting to note that slightly more than half of the respondents felt that it is negative whereas 27.58 per cent opined as positive.

NATURE OF BUREAUCRACY

There is a general agreement in the fact that India opted for a neutral rule-bound Weberian bureaucracy during the post independence period. In the initial decades of independent India not much problem was perceived in the administrative working. But later problems started cropping up and during Mrs. Gandhi tenure, the realisation became more vocal. This called for a re-examination of the nature of bureaucracy. In this context, it could be mentioned that a concept of committed bureaucracy emerged on the national scene as against the neutral bureaucracy. But this could not cut much ice.

We also wanted to assess the opinion of the officers regarding the nature of bureaucracy. We asked them to comment as to whether the bureaucracy should be caste neutral. It is interesting to note that a majority of respondents have opined that it should be caste neutral and very few reported that it should not be caste neutral.

It has been observed earlier that in Uttar Pradesh the role of caste in societal process is predominant. We wanted to assess the role of caste in the administrative process. The response pattern received shows that majority of the respondents have expressed their opinion that it has increased. About 27.58 percent only felt that it has decreased. It can be observed here that while the respondents have favoured a caste neutral bureaucracy, they find that the role of caste has increased.

In a multiparty system different political parties participate in the political and administrative process. Also these parties have varying political culture which influence the administrative process. We tried to ascertain from the respondents as to whether the extent of interference varies with different political parties. It has positively been reported that yes, it varies with political parties which establishes the linkages of political culture and administration. The respondents have also expressed that different political parties have different political culture.

Since the political culture influences the administrative process, we wanted to ascertain from the respondents about various considerations in postings. The considerations mentioned are political, merit, caste and money. The respondents have quoted most of these options in their response. It should not be alarming to find money as a consideration or caste as a consideration in posting. Only difference in the option is that of the gradation.

The domination of the caste factor in the administrative process has also been confirmed by the respondents. A considerable percentage of respondents (32.76 per cent) have expressed that it largely dominates the administrative process whereas about 64 percent felt that it sometimes affects the administrative process. It only goes to show that respondents have felt the domination of caste factor in the administrative decision making process, variation is only in the degree of the domination.

The above analysis is based on the data collected through the structured questionnaire. It is now proposed to give a brief account based on the observation during the course of the field work. It may be mentioned that during the course of field survey, the three major concerns were – **the service, the work** and **the attitude** of the civil servants. Our observations were guided by these three major concerns.

It has generally been observed that there has been an increasing interference in the day-to-day administration. The officers confessed in private that due to excessive politisation of the administrative system, the efficiency and governance has been suffering. Over the years, the political leadership is trying to indulge in routine administrative matters which many a times result in role crisis. This role crisis is observed in the decision making process and governance.

It has often been seen that honest upright officers feel neglected and isolated in the present rut of administration. Their motivational level too seems to be rather low. They remain reluctant to offer frank advice, as the same may not be liked by the political masters. They, therefore, opt for the path of least resistance although this may not be in the best of interest of the state and the people.

Different political parties have different political culture. The respondents had earlier reported about their variation in cultural pattern. Depending on these the extent of political interference in administrative matter has been observed. The old established political parties indulge less in administrative matters while the new ones believe in extensive interference in these matters.

The Secretariat of the State is the apex decision making body. It is necessary for the Secretariat to have an intimate knowledge of national politics, programmes and perspectives. How far they have knowledge is a matter of opinion as sometimes we have observed the lack of these during the course of our field survey.

It can be noted that during the last few years the State is being run by a coalition of parties. This is a new experiment in the political life of the country as well as the state of Uttar Pradesh. It has been observed that due to its varying nature, it has affected the decision making process adversely. The reason being that in the process of governance, short time goals, political and personal, become more important than long term societal goals. This has affected the perception as well as the attitude of politicians and the civil servants. There seems to be a continuous struggle for power which cause damage to the public system.

It can, therefore, be observed that caste as a factor has a role to play in the decision making process alongwith other extraneous factors. It has influenced the attitudes of the civil servants and at times causes a divide between the different groups of civil servants. The civil service no longer look a homogeneous group of people, it is a rather heterogeneous body of people representing different social ethos. These are the observations based on data and information provided by the respondents.

It may be noted here that the state of Uttar Pradesh for the last almost fifteen years has witnessed several chief ministers who had their own value premises and orientations as well as a distinct style and focus. We have already made an observation about these aspects in the earlier part of the book. We would like to briefly make a concluding observation for re-enforcing the point that the caste has gone deep into the body politic of the state and society. Under such circumstances we find Deen Dayal Upadhyay, Ambedkar, Shahu Ji Maharaj, Lohia to dominate the decision-making and the governance at the policy level. We thus, notice the caste factor influencing the postings, transfers, promotions which consequently influences the quality of governance. The public issues become no issues and the governance becomes casualty.

In view of the issues involved and the observations made, there seems a need to re-assess our priorities in the decision making and the governance, if we believe in a non-discriminatory decision making process. Let the governance by for the general mass of the people. Ignoring vital public interest and welfare could be counter productive which may ultimately damage the political system itself.

Annexure – I

Questionnaire for Officials

PART – A

I. GENERAL BACKGROUND

1. Name : ....
2. Age : ....
3. Sex : ....
4. Educational Qualification : ....
5. Service Cadre : ....
6. Present Posting as : ....
7. Length of Present Posting : ....
8. Marietal Status : ....
9. Family Size : ....
10. Religion : ....
11. Caste : ....
12. Monthly Income : ....

PART – B

II. SERVICE RELATED

1. How long have you been
   - (a) At the Secretariat
   - (b) At the Field

2. Which posting you enjoyed most?
   Secretariat / Field

3. Which place you find working difficult?
   Secretariat / Field

4. Which place the political interference is more?
   Secretariat / Field

5. How do you reach to these political interference?
   Don't Pay Attention/ Consider as Normal in the System/ Yield to them.

6. How often you have taken recourse to approach for favour in your service?
   Often/ Sometimes/ Never

7. How often your caste people approach you for help?
   Very Often/ Sometimes/ Never

8. How do you reach to this approach?
   Positive/ Negative

9. Do you think in a caste ridden society, the bureaucracy be caste neutral?
   Yes/ No

10. In your opinion, should the bureaucracy be caste neutral?
    Yes/ No

11. In your opinion the role of caste in socio-political process over the years has:
    Increased/ Decreased/ No Change

12. How far do you think that plays a positive role in bringing social change?
    Largely/ To Some Extent/ Not at All

III. WORK RELATED

1. What is the present nature of work you are concerned with?

2. How far it involves the interest of the people?
   Largely/ To Some Extent/
   Not at All

3. How far are you satisfied with the present work?
   Largely/ To Some Extent/
   Not at All

4. Do you think that the extent of interference varies with different political parties?
   Yes/ No

5. Do you think different political parties have different political culture?
   Yes/ No

6. Do you think that interference increases/ decreased with following political parties in power?

   Please tick against the answer.

Party	**Increases**	**Decreases**
Congress	( )	( )
B.J.P.	( )	( )
S.P.	( )	( )
B.S.P.	( )	( )
J.D.	( )	( )
Others (Name the party)	( )	( )

7. During which political party rule the bureaucracy seems as more comfortable: (Please tick against the answer)

   Please tick against the answer.

Party	**Answer**
Congress	( )
B.J.P.	( )
S.P.	( )
B.S.P.	( )

J.D. ( )

Others ( )

(Name the party)

8. How far do you think the social background of politician is reflected in their behaviour?

Largely/ To Some Extent/ Not at All

9. How far do you think the social background is reflected in their behaviour?

Largely/ To Some Extent/ Not at All

10. In your opinion what factors affect the work culture of civil servants?

Largely/ To Some Extent/ Not at All

11. How far are you satisfied with present culture?

Largely/ To Some Extent/ Not at All

12. In view of the present day needs, how far do you think there is a need to change the present work culture?

Largely/ To Some Extent/ Not at All

13. In case of need, what measures do you suggest for effecting that change?

14. In your opinion, on which of the following considerations generally postings are based?

1. Political Considerations
2. Caste Considerations
3. Merit Considerations
4. Money Considerations
5. Other Considerations (Please Specify)

IV. ATTITUDINAL QUESTIONS

1. Do you belong Urban/ Rural Area?
2. How much land do you own in your village?
3. Which of the following place your schooling was done?

   At Village/ Municipal School/ Public School/ Govt. School/ Private School
4. Where does your children go to?

   Public School/ Govt. School/ Private School
5. How far do you think reservation has helped in restoring merit?

   Largely/ To Some Extent/ Not at All
6. Which of the following considerations guide administrative decision?

   (a) Political

   (b) Social

   (c) Rules on the subjects

   (d) Caste

   (e) Any Other (Please Specify)
7. It is generally seen that caste factor is predominant in socio-political process, how far do you think it dominates the administrative process as well?

   Largely/ To Some Extent/ Not at All
8. How far do you think caste as a factor administration is going to help solve problem in the country?

   Largely/ To Some Extent/ Not at All
9. The bureaucracy is now heterogeneous, how far do you

think it reflects change in attitudes of different groups within the bureaucracy?

Largely/ To Some Extent/
Not at All

10. How far do you think divergent attitudes will help in promoting social cohesion?

Largely/ To Some Extent/
Not at All

11. How do you mix up with following caste people?

Category	With Reservation	Without Reservation
Colleagues	( )	( )
Subordinate	( )	( )
Other Service Cadre	( )	( )
Other Groups	( )	( )

12. How far do you think the attitudes of the bureaucracy is helpful in accelerating growth and change?

Largely/ To Some Extent/
Not at All

13. How far do you think the heterogeneous character of the society has affected the administrative decision making?

Largely/ To Some Extent/
Not at All

14. How far do you think the alleged elitist nature of administrative has affected the development process?

Largely/ To Some Extent/
Not at All

15. How far do you think the heterogeneous nature of society is also reflected in the bureaucracy?

Largely/ To Some Extent/
Not at All

16. How far do you think it is the cause of development and growth?

Largely/ To Some Extent/
Not at All

Annexure - II
TABLES

Table 5.1
POSTING-WISE PERCENTAGE WHICH WAS ENJOYED MOST

Posting	*Percentage*
Secretariat	43.10
Field	55.17
No Response	01.73

Table 5.2
POSTING WISE WORKING DIFFICULT

Posting	*Percentage*
Secretariat	31.04
Field	55.17
No where	05.17
No Response	08.62

Table 5.3

POLITICAL INTERFERENCE IS MORE

Place	*Percentage*
Secretariat	29.31
Field	58.62
No where	06.90
No Response	05.17

Table 5.4

REACTION TO POLITICAL INTERFERENCE

Responses	*Percentage*
Don't Pay Attention	03.45
Consider As Normal	77.58
Yield to Them	01.73
No Response	17.24

Table 5.5

RECOURSE TO APPROACH FOR FAVOUR IN THE SERVICE

Response	*Percentage*
Often	06.98
Sometimes	44.83
Never	44.83
No Response	03.45

Table 5.6

CASTE PEOPLE APPROACH FOR HELP

Responses	*Percentage*
Very Often	13.79
Sometime	58.62
Never	24.14
No Response	03.45

Table 5.7

REACTION OF THE APPROACH

Response	*Percentage*
Positive	27.58
Negatives	51.73
Others	12.07
No Response	08.62

Table 5.8

SHOULD THE BUREAUCRACY BE CASTE-NEUTRAL

Responses	***Percentage***
Yes	81.03
No	17.24
No Response	01.73

Table 5.9

THE ROLE OF THE CASTE

Response	*Percentage*
Increased	67.24
Decreased	27.58
No Response	05.18

Table 5.10

ROLE IN SOCIAL CHANGE

Responses	*Percentage*
Largely	15.52
To some Extent	50.00
Not at All	31.03
No Response	03.45

Table 5.11

EXTENT OF INVOLVEMENT WITH PUBLIC INTEREST

Response	*Percentage*
Largely	36.20
To some Extent	46.55
Not at All	15.52
No Response	01.73

Table 5.12

SATISFACTION LEVEL WITH THE PRESENT WORK

Response	*Percentage*
Largely	29.31
To some Extent	62.07
No Response	08.62

Table 5.13

EXTENT OF INTERFERENCE VARIES WITH DIFFERENT POLITICAL PARTIES

Responses	*Percentage*
Yes	72.41
No	25.86
No Response	01.73

Table 5.14

DIFFERENT POLITICAL PARTIES HAVE DIFFERENT POLITICAL CULTURE

Responses	*Percentage*
Yes	84.48
No	15.52

Table 5.15

SOCIAL BACKGROUND IN BEHAVIOUR

Response	*Percentage*
Largely	51.73
To some Extent	44.82
No Response	03.45

Table 5.16

REFLECTION OF SOCIAL BACKGROUND IN BEHAVIOUR

Responses	*Percentage*
Largely	36.22
To some Extent	48.27
Not at All	03.45
No Response	12.06

Table 5.17

SATISFACTION WITH PRESENT WORK CULTURE

Response	*Percentage*
Largely	08.63
To some Extent	68.97
Not at All	20.68
No Response	01.73

Table 5.18

NEED TO CHANGE PRESENT WORK CULTURE

Responses	*Percentage*
Largely	60.34
To some Extent	34.49
Not at All	05.17

Table 5.19

CONSIDERATION IN POSTING

Response	*Percentage*
Political	25.86
Caste	10.34
Merit	31.04
Money	17.24
Others	01.73
No Response	13.79

Table 5.20

MODE OF SCHOOLING OF THE RESPONDENTS

Responses	*Percentage*
Village School	17.24
Municipal School	13.79
Public School	17.24
Government School	39.66
Private School	10.34
No Response	01.73

Table 5.21

SCHOOLING OF CHILDREN

Responses	*Percentage*
Public School	27.59
Government School	25.86
Private School	39.66
No Response	06.89

Table 5.22

RESERVATION HELPED IN RESTORING MERIT

Responses	*Percentage*
Largely	13.79
To some Extent	50.00
Not at All	34.48
No Response	01.73

Table 5.23

CONSIDERATIONS THAT GUIDE YOUR ADMINISTRATIVE DECISION

Responses	*Percentage*
Political	10.34
Social	10.34
Rules on the Subject	68.97
Caste	05.17
Others	01.73
No Response	03.45

Table 5.24

DOMINATION OF CASTE FACTOR IN ADMINISTRATION

Responses	*Percentage*
Largely	32.76
To some Extent	63.79
Not at All	03.45

Table 5.25

CASTE FACTOR HELPS IN SOLVING THE PROBLEM

Responses	*Percentage*
Largely	06.89
To some Extent	31.04
Not at All	62.07

Table 5.26

WHETHER HETEROGENEITY GETS REFLECTED IN CHANGE IN ATTITUDES

Responses	*Percentage*
Largely	18.96
To some Extent	72.41
Not at All	08.63

Table 5.27

DIVERGENT ATTITUDES HELP TO PROMOTE SOCIAL COHESION

Responses	*Percentage*
Largely	18.96
To some Extent	62.08
Not at All	12.07
No Response	06.89

Table 5.28

ATTITUDE IN BUREAUCRACY IN ACCELERATING GROWTH AND CHANGE

Responses	*Percentage*
Largely	32.75
To some Extent	51.73
Not at All	15.52

Table 5.29

HETEROGENEOUS CHARACTER HAS AFFECTED THE DECISION MAKING

Responses	*Percentage*
Largely	27.58
To some Extent	68.97
Not at All	03.45

Table 5.30

ALLEGED ELITIST NATURE HAS AFFECTED DEVELOPMENT PROCESS

Responses	*Percentage*
Largely	27.58
To some Extent	68.97
Not at All	03.45

Table 5.31

HETEROGENEOUS NATURE OF SOCIETY IS REFLECTED IN BUREAUCRACY

Responses	*Percentage*
Largely	22.41
To some Extent	68.97
Not at All	06.89
No Response	01.73

Table 5.32

HETEROGENEITY HELPING THE CAUSE OF DEVELOPMENT AND GROWTH

Responses	*Percentage*
Largely	18.96
To some Extent	25.18
Not at All	24.13
No Response	01.73

□□□